T0249845

17. Immune Regulation: Evolution and Biological Significance
Edited by Laurens N. Ruben and M. Eric Gershwin

18. Tumor Immunity in Prognosis: The Role of Mononuclear Cell Infiltration
Edited by Stephen Haskill

19. Immunopharmacology and the Regulation of Leukocyte Function
Edited by David R. Webb

20. Pathogenesis and Immunology of Treponemal Infection
Edited by Ronald F. Schell and Daniel M. Musher

21. Macrophage-Mediated Antibody-Dependent Cellular Cytotoxicity
Edited by Hillel S. Koren

22. Molecular Immunology: A Textbook
Edited by M. Zouhair Atassi, Carel J. van Oss, and Darryl R. Absolom

23. Monoclonal Antibodies and Cancer
Edited by George L. Wright, Jr.

24. Stress, Immunity, and Aging
Edited by Edwin L. Cooper

25. Immune Modulation Agents and Their Mechanisms
Edited by Richard L. Fenichel and Michael A. Chirigos

26. Mononuclear Phagocyte Biology
Edited by Alvin Volkman

27. The Lactoperoxidase System: Chemistry and Biological Significance
Edited by Kenneth M. Pruitt and Jorma O. Tenovuo

28. Introduction to Medical Immunology
Edited by Gabriel Virella, Jean-Michel Goust, H. Hugh Fudenberg, and Christian C. Patrick

29. Handbook of Food Allergies
Edited by James C. Breneman

30. Human Hybridomas: Diagnostic and Therapeutic Applications
Edited by Anthony J. Strelkauskas

31. Aging and the Immune Response: Cellular and Humoral Aspects
Edited by Edmond A. Goidl

32. Complications of Organ Transplantation
Edited by Luis H. Toledo-Pereyra

33. Monoclonal Antibody Production Techniques and Applications
Edited by Lawrence B. Schook

34. Fundamentals of Receptor Molecular Biology
Donald F. H. Wallach

35. Recombinant Lymphokines and Their Receptors
Edited by Steven Gillis

36. Immunology of the Male Reproductive Organs
Edited by Pierre Luigi Bigazzi

37. The Lymphocyte: Structure and Function
Edited by John J. Marchalonis

38. Differentiation Antigens in Lymphohemopoietic Tissues
Edited by Masayuki Miyasaka and Zdenek Trnka

39. Cancer Diagnosis In Vitro Using Monoclonal Antibodies
Edited by Herbert Z. Kupchik

40. Biological Response Modifiers and Cancer Therapy
Edited by J. W. Chiao

41. Cellular Oncogene Activation
Edited by George Klein

42. Interferon and Nonviral Pathogens
Edited by Gerald I. Byrne and Jenifer Turco

43. Human Immunogenetics: Basic Principles and Clinical Relevance
Edited by Stephen D. Litwin, with David W. Scott, Lorraine Flaherty, Ralph A. Reisfeld, and Donald M. Marcus

44. AIDS: Pathogenesis and Treatment
Edited by Jay A. Levy

45. Cell Surface Antigen Thy-1: Immunology, Neurology, and Therapeutic Applications
Edited by Arnold E. Reif and Michael Schlesinger

46. Immune Mechanisms in Cutaneous Disease
Edited by David A. Norris

47. Immunology of Fungal Diseases
Edited by Edouard Kurstak

48. Adoptive Cellular Immunotherapy of Cancer
Edited by Henry C. Stevenson

Additional Volumes in Preparation

ADOPTIVE CELLULAR IMMUNOTHERAPY OF CANCER

IMMUNOLOGY SERIES

1. Mechanisms in Allergy: Reagin-Mediated Hypersensitivity
   Edited by Lawrence Goodfriend, Alec Sehon and Robert P. Orange
2. Immunopathology: Methods and Techniques
   Edited by Theodore P. Zacharia and Sidney S. Breese, Jr.
3. Immunity and Cancer in Man: An Introduction
   Edited by Arnold E. Reif
4. *Bordetella pertussis:* Immunological and Other Biological Activities
   J.J. Munoz and R.K. Bergman
5. The Lymphocyte: Structure and Function (in two parts)
   Edited by John J. Marchalonis
6. Immunology of Receptors
   Edited by B. Cinader
7. Immediate Hypersensitivity: Modern Concepts and Development
   Edited by Michael K. Bach
8. Theoretical Immunology
   Edited by George I. Bell, Alan S. Perelson, and George H. Pimbley, Jr.
9. Immunodiagnosis of Cancer (in two parts)
   Edited by Ronald B. Herberman and K. Robert McIntire
10. Immunologically Mediated Renal Diseases: Criteria for Diagnosis and Treatment
    Edited by Robert T. McCluskey and Giuseppe A. Andres
11. Clinical Immunotherapy
    Edited by Albert F. LoBuglio
12. Mechanisms of Immunity to Virus-Induced Tumors
    Edited by John W. Blasecki
13. Manual of Macrophage Methodology: Collection, Characterization, and Function
    Edited by Herbert B. Herscowitz, Howard T. Holden, Joseph A. Bellanti, and Abdul Ghaffar
14. Suppressor Cells in Human Disease
    Edited by James S. Goodwin
15. Immunological Aspects of Aging
    Edited by Diego Segre and Lester Smith
16. Cellular and Molecular Mechanisms of Immunologic Tolerance
    Edited by Tomáš Hraba and Milan Hašek

ADOPTIVE CELLULAR IMMUNOTHERAPY OF CANCER

EDITED BY

HENRY C. STEVENSON

Division of Cancer Treatment
National Cancer Institute
National Institutes of Health
Bethesda, Maryland

Technical Editor

DEVERA G. SCHOENBERG

National Institutes of Health
Bethesda, Maryland

MARCEL DEKKER, INC.   NEW YORK • BASEL

Library of Congress Cataloging-in-Publication Data

Adoptive cellular immunotherapy of cancer / Henry C. Stevenson, editor.
p. cm.–(Immunology series)
Includes index.
ISBN 0-8247-8111-2
1. Cancer–Immunotherapy. 2. Cellular therapy. 3. Immunocompetent cells–Therapeutic use. I. Stevenson, Henry C., [date] II. Series.
[DNLM: 1. Immunity, Cellular. 2. Immunotherapy–methods. 3. Neoplasms–therapy. W1 IM53K / QZ 266 A239]
RC271.I45A33 1989
616.99'4061–dc20
DNLM/DLC
for Library of Congress 89-1418
CIP

This book is printed on acid-free paper

MARCEL DEKKER, INC.
270 Madison Avenue, New York, New York 10016

Current printing (last digit):
10 9 8 7 6 5 4 3 2 1

PRINTED IN THE UNITED STATES OF AMERICA

Preface

The 1980s has ushered in a host of scientific achievements that have captured the imagination of cancer researchers and practitioners worldwide. Accompanying this explosion of new insights and information has come the responsibility for processing, condensing, and disseminating clinically applicable material to clinicians and their co-workers in the cancer patient care arena. Clearly, the biotherapy of cancer represents one of the significant breakthroughs in oncology of this decade and has been designated the "fourth modality" of cancer therapy, joining surgery, radiation therapy, and chemotherapy. The newly emerging adoptive cellular immunotherapies (ACI) of cancer represent the most complicated and potentially most powerful cancer biotherapies developed to date.

This ACI volume was prepared with cancer biotherapy clinicians and their co-workers in mind. The coauthors of this text have thoughtfully gleaned novel insights on ACI and packaged them in a format that should prove understandable and useful to ACI teams in the field. Following an overview of human immune system function and the mechanisms by which ACI harnesses this activity, a brief glossary of useful terms and concepts is presented in the introductory overview chapter (Chap. 1). Next, there are five chapters (2-6) on lymphokine activated killer lymphocytes (LAK) that provide a preclinical and clinical summary of this novel human immunotherapy phenomenon. Two chapters (7 and 8) review our preclinical and clinical understandings of human activated killer monocytes (AKM). Two additional chapters (9 and 10) review tumor-infiltrating lymphocytes (TIL) and tumor-derived activated cells (TDAC). Three chapters (11-13) round out the text by focusing on the clinical and technical problems of the multidisciplinary approach to ACI therapy; these issues are examined at the level of cytapheresis, in vitro technology, and nursing/technical support. Team effort and active participation at all levels in the execution of ACI trials are emphasized. Chapter 14 provides an evolutionary vision of the numerous exciting developments in ACI therapy that we are likely to encounter over the next few years.

Adoptive Cellular Immunotherapy of Cancer was written with the hope that it might provide an easily understandable framework for the myriad of ACI preclinical and clinical observations available in the literature. Moreover, every attempt has been made to provide some information of interest to each of the groups of professionals that constitute the modern ACI multidisciplinary team. We hope that as this volume promotes greater scientific capability in each individual member of the ACI team, it will also promote greater group cooperation and enhanced scientific productivity of the ACI team as a whole: It is to this unique assemblage of caring individuals that this book is dedicated.

Henry C. Stevenson

Contributors

Suzanne Beckner, M.D. Biological Response Modifiers Program, National Cancer Institute, Frederick, Maryland

J. Bob Blacklock, M.D. Neurosurgery Section, Department of Head and Neck Surgery, M. D. Anderson Hospital, The University of Texas System Cancer Center, Houston, Texas

Edwin Burgstaler, M.D. Mayo Clinic Blood Bank, Mayo Clinic, Rochester, Minnesota

Jeffrey Clark, M.D. Clinical Investigations Branch, Biological Response Modifiers Program, National Cancer Institute, Frederick, Maryland

Margaret Farrell, R.N. Biological Response Modifiers Program, National Cancer Institute, Frederick, Maryland

Colleen S. Friddell Cellular Immunology Section, Biotherapeutics, Inc., Franklin, Tennessee

Elizabeth A. Grimm, M.D. Departments of Tumor Biology and

Surgery, M.D. Anderson Hospital, The University of Texas System Cancer Center, Houston, Texas

D. S. Heo, M.D. Departments of Pathology and Medicine, University of Pittsburgh School of Medicine and Pittsburgh Cancer Institute, Pittsburgh, Pennsylvania

Ronald B. Herberman, M.D. Departments of Pathology and Medicine, University of Pittsburgh School of Medicine and Pittsburgh Cancer Institute, Pittsburgh, Pennsylvania

Jeane Hester, M.D. Department of Transfusion Medicine, M.D. Anderson Hospital, The University of Texas System Cancer Center, Houston, Texas

Martin R. Jadus, M.D. Cellular Immunology Section, Biotherapeutics, Inc., Franklin, Tennessee

Harvey G. Klein, M.D. Department of Transfusion Medicine, Warren G. Magnuson Clinical Center, National Institutes of Health, Bethesda, Maryland

Leocadio V. Lacerna, M.D. Biological Response Modifiers Program, National Cancer Institute, Frederick, Maryland

Walter M. Lewko Cellular Immunology Section, Biotherapeutics, Inc., Franklin, Tennessee

Dan L. Longo, M.D. Biological Response Modifiers Program, National Cancer Institute, Frederick, Maryland

James R. Maleckar, M.D. Cellular Immunology Section, Biotherapeutics, Inc., Franklin, Tennessee

Paul J. Miller Biological Response Modifiers Program, National Cancer Institute, Frederick, Maryland

Robert K. Oldham Biological Therapy Institute, Franklin, Tennessee

Laurie B. Owen-Schaub, M.D. Department of Tumor Biology, M.D. Anderson Hospital, The University of Texas System Cancer Center, Houston, Texas

Alvaro A. Pineda, M.D. Mayo Clinic Blood Bank, Mayo Clinic, Rochester, Minnesota

Steven Rudnick Biological Response Modifiers Program, National Cancer Institute, Frederick, Maryland

Hiroto Shinomiya, M.D. Biological Response Modifiers Program, National Cancer Institute, Frederick, Maryland

Mizuho Shinomiya, M.D. Biological Response Modifiers Program, National Cancer Institute, Frederick, Maryland

Irena J. Sniecinski, M.D. Blood Transfusion Services, City of Hope, National Medical Center, Duarte, California

G. W. Stevenson, M. D. Pediatric Anesthesia Department, Northwestern University Children's Hospital, Chicago, Illinois

Henry C. Stevenson, M. D. Division of Cancer Research, National Cancer Institute, Bethesda, Maryland

Paul H. Sugarbaker, M.D. Department of Surgical Oncology, The Windship Clinic, Emory University Medical Center, Atlanta, Georgia

S. Takagi, M.D. Departments of Pathology and Medicine, University of Pittsburgh School of Medicine and Pittsburgh Cancer Institute, Pittsburgh, Pennsylvania

William H. West, M.D. Biotherapeutics, Inc., Memphis, Tennessee

Theresa L. Whiteside, M.D. Department of Pathology and Medicine, University of Pittsburgh School of Medicine and Pittsburgh Cancer Institute, Pittsburgh, Pennsylvania

John R. Yannelli, M.D. Cellular Immunology Section, Biotherapeutics, Inc., Franklin, Tennessee

Contents

ADOPTIVE CELLULAR IMMUNOTHERAPY OF CANCER

1

Adoptive Cellular Immunotherapy of Cancer: An Overview

HENRY C. STEVENSON
National Cancer Institute, Bethesda, Maryland

A variety of distinct strategies to boost the immune system function of the cancer patient in an attempt to treat malignancy have been developed over the past several decades (1-6). It is clear from reviewing the immunological research of the past 100 years that the human immune system is capable of fulfilling its visualized objectives of eliminating nonself invaders from the body; this includes such distinct life forms as viruses, bacteria, and fungi. As will be detailed in this text, compelling evidence has recently been accumulated to indicate that host cells that have undergone malignant transformation not only behave as nonself invaders (attempting to take over the rest of the host), but they also bear chemical markers of malignant transformation (tumor antigens) that allow these cells to be "recognized" by the immune system; such malignantly transformed cells can also be destroyed by host immune system cells. The decade of the 1980s has focused extensively on the development of a new generation of cancer biotherapies—new immunotherapeutic techniques designed to "upregulate" human immune system function in an attempt to irradicate malignant cells from cancer patients.

Current biotherapy approaches to cancer treatment are diagramed in Figure 1. From a functional perspective, three different

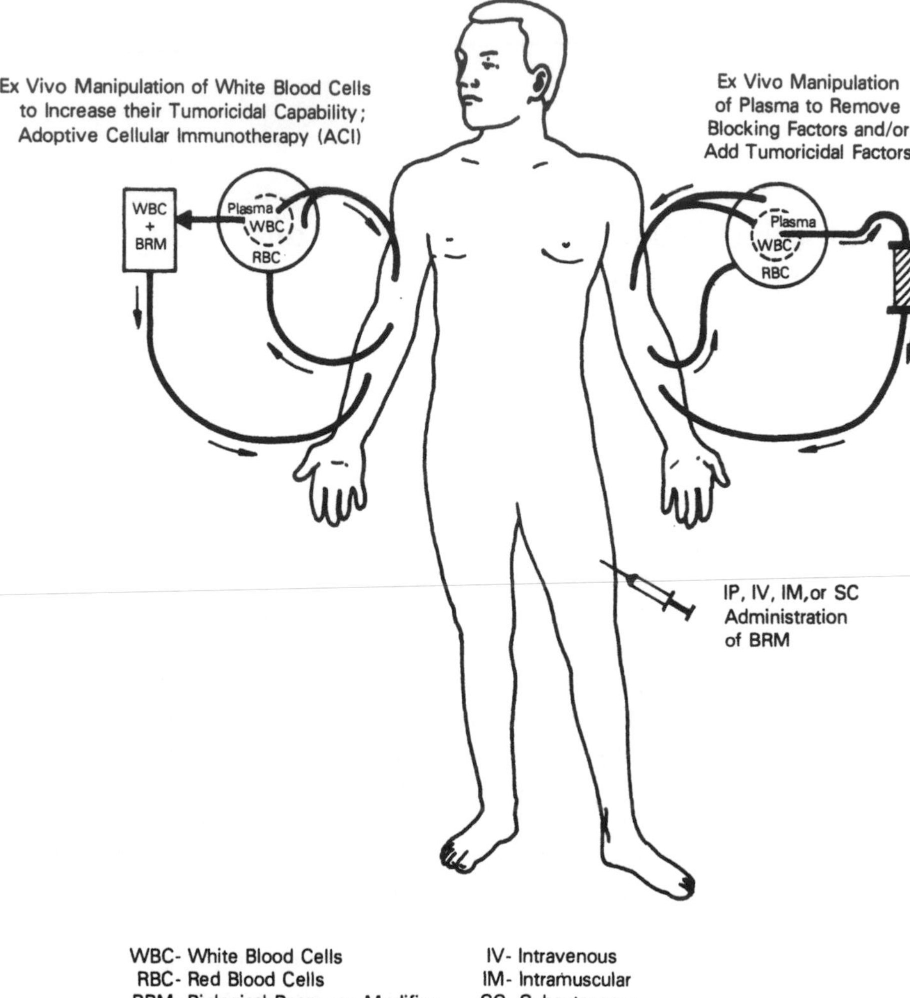

Figure 1 Representation of the functional mechanics involved in the performance of the three major types of cancer biotherapy: (a) Adoptive cellular immunotherapy (ACI), (b) ex vivo manipulation of patient plasma to remove blocking factors or add tumoricidal factors, and (c) in vivo administration of BRM.

types of biotherapy research strategies are now being explored. One strategy involves the ex vivo manipulation of the cancer patient's plasma in an effort to remove blocking factors or to add tumoricidal factors to the plasma before the plasma reinfusion. Such therapies usually require a centrifugation separation step to remove plasma from the other blood elements, followed by passage of the plasma over a separate column device such as is employed in staphyloccocal protein A column therapy (7). A more frequently employed biotherapy approach involves the direct in vivo treatment of cancer patients with biological response modifiers (BRM) in an attempt to utilize these agents to directly stimulate the failing immune system within the cancer patient's body. Such BRM include the interferons (IFN), the interleukins (IL), colony-stimulating factors (CSF), monoclonal antibodies (MAb) and other immunopotentiating agents, such as muramyldipeptide (MDP) and *Corynebacterium parvum* (*C. parvum*) (4,8,9). (More details about certain of these BRMs will follow in this chapter.) A final functional classification of current biotherapy focuses on the ex vivo activation of cancer patient's leukocytes with BRM in an attempt to expand their numbers or to increase their antitumor capabilities, or both; this approach has been termed adoptive cellular immunotherapy (ACI). Adoptive cellular immunotherapy is usually initiated with the performance of a cytapheresis procedure to separate the white blood cells (WBC) from the plasma and the red blood cells (RBC); the plasma and RBC are returned to the patient immediately; alternatively, the WBC may be isolated from patient tumor specimens directly. The WBC undergo further processing, including their incubation with BRM designed to augment or to increase their tumor-killing capabilities. These activated WBC are then reinfused back into the patient. The purpose of this volume is to present our current understandings of the mechanics and mechanisms of action of the ACI protocols currently being performed.

THE ANTITUMOR ACTIVITIES OF THE HUMAN IMMUNE SYSTEM

Within the past two decades, we have gleaned many insights into the overall operation of the human immune system and the potential applicability of this system to the treatment of cancer. As shown in Figure 2, the immune system mechanisms that are felt to be operative in the elimination of tumor cells from the body are rather straightforward. There are four basic cell types whose function has been associated with antitumor cell immunity. The B-lymphocytes secrete protein factors [immunoglobulins (Ig)] into the blood plasma, which have the capability of identifying and binding to targets that bear specific molecular markers that they can recognize. From a functional perspective, each individual possesses enough different types (clones) of B lymphocytes (and thus enough different Ig types) to identify most nonself invaders; this specific component of humoral (plasma-borne) immunity is a potent mechanism for identifying and labeling nonself invaders. It is noteworthy, however, that Ig-labeled target cells, in and of themselves, perform in a relatively unimpeded fashion; that is, additional factors are required for Ig-coated targets (such as tumor cells) to be destroyed. The two dominant mechanisms whereby Ig-coated targets are cleared from the body are through the complement cascade (the nonspecific component of humoral immunity) and the antibody-dependent cell-mediated cytotoxicity (ADCC) mechanisms (cellular mechanisms reviewed below). The complement proteins are a series of proteins that exist in an inactive state in the blood plasma under normal conditions. However, upon encountering an immunoglobulin (IgG or IgM)-coated target, these plasma proteins assemble into a trocarlike mechanism that is capable of lysing nonself invader cells (including tumor cells). The complement proteins are chiefly secreted by the mononuclear phagocyte series (blood monocytes). Since the advent of MAb technology (4), great attention has been focused on the protein products of B lymphocytes and the application of these Ig proteins to the treatment of cancer. For this reason, neither the Ig proteins nor their cellular factories (the B lymphocytes) will be discussed further in this volume, except to mention that the B lymphocytes are subject to activation and regulation by a variety of protein signal molecules—the BRM–chemotherapy agents, and other chemicals that can influence their function. These modulating agents will be reviewed relative to the other leukocyte subsets that participate in the overall immune response to cancer.

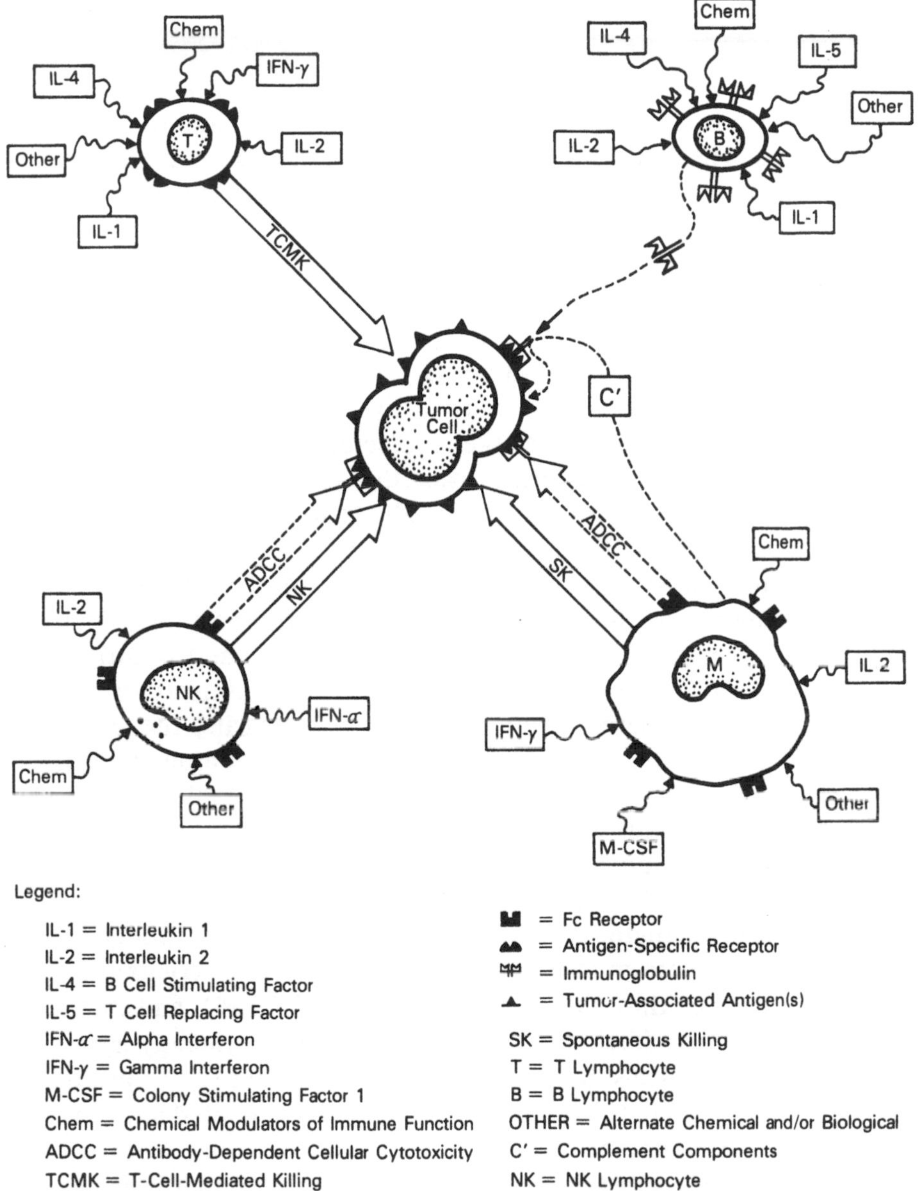

Figure 2 Representation of the potential mechanisms of action whereby B, T, and NK lymphocytes and mononuclear phagocytes (monocytes and macrophages) may eliminate tumor cells.

The two remaining lymphocyte subsets have been harnessed for ACI. They are the natural killer (NK) lymphocytes and the T lymphocytes. The T lymphocytes exist in clones as do the B lymphocytes; in contrast with B lymphocytes, they do not function by synthesizing proteins for transport through the blood stream to mediate tumor cell killing but, rather, they physically travel to the site of tumor cell invasion and locally arrange for the destruction of the tumor cells. As shown in Figure 2, T lymphocytes possess antigen-specific receptors, and each T-lymphocyte clone has the capacity to recognize (and potentially destroy) a tumor cell carrying complementary marker molecules on its membrane (tumor-associated antigens). Given the total spectrum of T-lymphocyte clones present in the body, it appears that each individual possesses the total range of T-lymphocyte clones required to eliminate most types of tumor cells (1). The T lymphocytes destroy tumor cells through their antigen-specific receptors. Such killing is termed T-cell–mediated killing (TCMK) and is antigen-specific; that is, each T-lymphocyte clone can eliminate only a very narrowly related group of tumor cells bearing the same tumor-associated antigen. Moreover, TCMK by T lymphocytes is restricted by the major histocompatibility (MHC) molecules; that is, tumor cell killing by T-lymphocyte clones requires identity between the MHC molecules found on the surface of the tumor cell and specific MHC receptors found on the surface of the T lymphocyte. T-cell–mediated killing can be modulated by a variety of influences. These include many BRMs, such as interleukin-1 (IL-1), interleukin-2 (IL-2) and interleukin-4 (IL-4), and interferon-γ (IFN-γ) (reviewed later), as well as chemotherapy agents and other chemicals. Some researchers believe that tumor-infiltrating lymphocyte (TIL) therapy is mediated by activated antigen-specific T-lymphocyte clones; however, this particular ACI topic is the subject of a substantial amount of controversy that is excellently summarized in Chapter 9. The developers of the tumor-derived activated cell (TDAC) therapy believe that they infuse activated antigen-specific T lymphocytes into their patients as reviewed in Chapter 10.

The natural killer (NK) lymphocyte is the third well-characterized lymphocyte subset that has antitumor cell reactivity. In contrast with the T lymphocyte, the NK lymphocyte bears no evidence of clonality; that is, a single type of cell appears to be capable of destroying a wide range of human tumors. This large-volume lymphocyte subset has distinctive granules in its cytoplasm and also has been termed the large granular lymphocyte (LGL; 10, 11). Natural killer lymphocytes have two distinct mechanisms for the destruction of tumor cells. One is a mechanism known as antibody-dependent cellular cytotoxicity (ADCC). This cytotoxicity occurs when special membrane receptors (Fc receptors) on the surface on the NK lymphocyte bind to the terminal (Fc) portion of Ig molecules (IgG) that have bound to the tumor-associated antigens found on the surface of tumor cells. The Ig molecule then functions as a ligand between the NK lymphocyte and the tumor cell and promotes the rapid destruction of the tumor cell. Alternatively, NK lymphocytes can kill tumor cells independent of Ig; this killing is termed natural killing (NK). The precise molecular marker that promotes natural killing is not now known; however, it appears to be a glycolipid molecule (ganglioside) that appears on virtually all malignantly transformed cells. The baseline natural-killing capability of NK cells can be upregulated by a variety of BRM including interferon-α (IFN-α) and IL-2 (reviewed later); it can also be modulated by chemotherapy agents and other chemicals. The lymphokine-activated killer lymphocyte (LAK) can be generated by upregulating the tumoricidal activity of NK lymphocytes by the in vitro incubation of this cell type with the BRM, IL-2; however other LAK precursor cells may exist. This issue is addressed in Chapter 2.

The third subset of WBC that can be harnessed for ACI trials is not a lymphocyte, but rather is from the mononuclear phagocyte series, chiefly in the form of blood monocytes. Like the NK lymphocytes, monocytes possess two independent mechanisms for destroying tumor targets. Monocytes are the dominant immune system cell possessing Fc receptors and, thus, they vigorously participate in the ADCC-mediated destruction of tumor cells. In addition, however, monocytes also are capable of destroying tumor targets independent of Ig, a process termed spontaneous cytotoxicity (SK). Like natural killing, SK appears to be triggered by a single class of tumor-associated molecules found on virtually all types of tumor cells. The SK-killing capability of monocytes against tumor cells can be upregulated by a variety of BRM; including IFN-γ, IL-2, and CSF (reviewed later), as well as certain chemotherapy agents and other chemicals. In vitro-activated cytotoxic monocytes are termed activated killer monocytes (AKM).

A substantial recent effort has focused on the question of "who is doing what" in ACI. Immunologists have devised a variety of techniques to characterize and identify individual leukocyte subsets to further their understandings of leukocyte physiology. Perhaps the most powerful tool for identifying these subsets (phenotyping) involves the use of MAbs. Individual MAbs are capable of recognizing distinct differentiation antigens found on specific subsets of leukocytes. A number of the molecules recognized by MAbs are shared widely among the leukocyte subsets; alternatively, certain molecules are found predominantly on only one type of leukocyte subset. The MAbs that recognize these specific leukocyte subset-differentiation antigens have proved quite helpful in the phenotyping process. Unfortunately, distinguishing between those MAbs that recognize relatively subset-specific differentiation molecules versus molecules widely shared among subsets has proven to be difficult. To add complexity to the problem, a variety of competing commercial sources of MAbs have given different proprietary names to their products, even though they may recognize the same molecule. Recently, conferences to resolve this nomenclature problem have been able to agree on a so-called cluster designation (CD) nomenclature for many of the molecules recognized by the current spectrum of commercial MAbs available. Table 1 provides a listing of some of the most commonly cited MAbs that react with T lymphocytes, monocytes, NK lymphocytes, and B lymphocytes; in addition, a MAb against the IL-2 receptor has also been identified. Each of the molecules felt to be recognized by these individual MAbs is listed according to its CD terminology, by the terminology utilized by the Ortho Corporation (which produces the OK series), by the Becton-

Table 1 Listing of Monoclonal Antibodies (MAbs) Recognizing Specific Molecules on Human Leukocyte Subsets (By CD, OK, Leu, and Other Nomenclatures)

Predominant cell type identified by MAbs	CD molecule identification	OK molecule identification	Leu molecule identification	Other nomenclature
T lymphocytes (sheep red blood cell receptor)	CD2	OKT11	Leu5b	T11
T lymphocytes	CD3	OKT3	Leu4	T3
Helper T lymphocytes	CD4	OKT4	Leu3a	T4
Suppressor/cytotoxic T lymphocytes	CD8	OKT8 OKT5	Leu2a Leu2b	T5; T8
Monocytes	CD15		LeuM1	My1
Monocytes			LeuM3	Mo2
Monocytes			LeuM5	KiM1
NK lymphocytes			Leu19	NKH-1
Fc receptor of NK lymphocytes and neutrophils	CD16		Leu11a–c	VEP13
B lymphocytes	CD19–22		Leu CR1 Leu12, 14, and 16	B1–B4
Interleukin-2 receptor	CD25		Interleukin-2 receptor	Tac

Table 2 Certain Frequently Cited Biological Response Modifiers (BRM); Their Chief Cell Source and Principal Activities

BRM	Cell source	Activities
IL-1 (interleukin-1; lymphocyte-activating factor)	Monocytes	Promotes early phases of the B and T lymphocyte activation processes
TNF-β (tumor necrosis factor)	Monocytes	Can directly kill certain types of tumor cells
IFN-α (alpha-interferon)	Monocytes	Activates NK lymphocytes; promotes HLA-DR expression on tumor cells; inhibits growth of certain tumor cells (hairy cell leukemia)
TGF-β (transforming growth factor-β)	Monocytes	Can promote tumor cell growth and down-regulate certain lymphocyte functions
G-CSF (granulocyte colony-stimulating factor)	Monocytes	Stimulates granulocyte differentiation
M-CSF (monocyte colony-stimulating factor)	Monocytes	Stimulates monocyte production; activates monocyte cytotoxic activity

IL-2 (interleukin-2; T-cell growth factor)	T Lymphocytes	Promotes activation of B, T, and NK lymphocytes; promotes proliferation of T and NK lymphocytes; enhances cytotoxic function of T and NK lymphocytes
IL-3 (interleukin-3; Multi-CSF)	T Lymphocytes	Promotes differentiation of platelets, eosinophils, basophils, neutrophils, monocytes, and RBCs
IL-4 (interleukin-4; B-cell-stimulating factor)	T Lymphocytes	Stimulates B-lymphocyte and cytotoxic T-lymphocyte activation
IL-5 (interleukin-5; T-cell-replacing factor)	T Lymphocytes	Promotes B-lymphocyte differentiation and immunoglobulin secretion
GM-CSF (granulocyte-monocyte colony-stimulating factor)	T Lymphocytes	Stimulates granulocyte and monocyte production; stimulates monocyte cytotoxic activity
IFN-γ (gamma-interferon)	T Lymphocytes	Inhibits growth of certain tumor cells; activates cytotoxic monocytes; activates T lymphocytes
TNF-α (lymphotoxin)	T Lymphocytes	Can kill certain types of tumor cells directly

Dickinson Corporation (which produces the Leu series), as well as other nomenclature systems currently available. This MAb listing is provided to the reader in the hope that (although cumbersome) the MAb terminology found throughout this text will not be overly confusing. As the reader continues reading in this volume and other scientific literature on ACI, one will see this terminology utilized variously and interchangeably: For example, a quick reference to Table 1 will allow the reader to determine which specific leukocyte subset is being referred to when a cell is said to be "CD3-positive and HNK1-negative" (a T lymphocyte) or "CD16-positive and LeuM5-negative" (an NK lymphocyte).

Another major area of current interest in cancer biotherapy is the new-generation BRM molecules that have been identified as having immunomodulatory properties. Table 2 summarizes a few of the most frequently cited BRMs; six BRMs that are chiefly produced by monocytes (monokines) are summarized first, followed by seven lymphocyte-derived BRMs (termed lymphokines). Each of these BRMs is known to be a distinct low-molecular-weight polypeptide molecule; we are now able to produce these proteins synthetically by recombinant DNA technology. Because most of these BRMs are in clinical trials as single agents (or shortly will be tested clinically), the reader is likely to encounter references to their activities, not only in this ACI volume, but also in other scientific reading on cancer biotherapy. Interferon-α, IL-2, and IFN-γ are currently being employed in ACI trials; TNF-β, M-CSF, and additional members of the interleukin (IL) family are expected to be employed shortly. Table 2 cites the abbreviated name of each BRM, the other names by which the factor is known, as well as the chief immunomodulatory function(s) that each BRM subserves.

GENERAL DESIGN OF ACI TRIALS

The general design of ACI trials is summarized in Figure 3. Patients with measurable metastatic malignancies that are refractory to standard therapies and who are still in good performance status are eligible for most ACI trials. Other entry criteria usually include

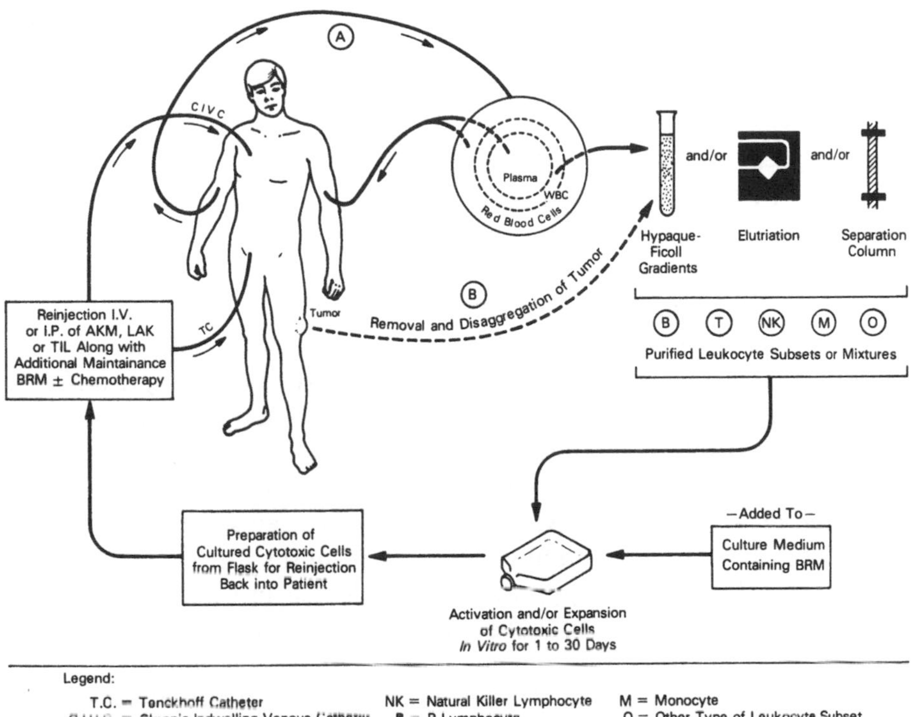

Figure 3 General schematic representation of the design of most adoptive cellular immunotherapy (ACI) protocols. Reprinted with permission from ZIB/SPS, Inc.

the capability of obtaining adequate venous access; the absence of significant hepatic, pulmonary, renal, infectious, or cardiovascular diseases; no recent (within 4 weeks) administration of potentially curative interventions (such as radiation therapy, chemotherapy, or biotherapy); and normal WBC counts; most protocols also exclude patients with central nervous system disease. The WBC are then harvested from the patients by one of two mechanisms: (a) For the lymphokine-activated killer (LAK) and the activated killer monocyte (AKM) protocols, leukocytes are harvested by cytapheresis; plasma and RBC are returned immediately to the patient. (b) For the tumor-infiltrating lymphocyte (TIL) and tumor-derived activated cell (TDAC) protocols, a tumor specimen is removed and the cells infiltrating the tumor are disaggregated, usually by a combination of mechanical and enzymatic means. The WBC harvested by either mechanism can be further purified by Hypaque–Ficoll (H–F) gradients, countercurrent centrifugal elutriation (CCE), column separation techniques, or a combination of these. The resulting cell preparation has varying degrees of purity for the desired leukocyte subset. For example, the LAK cell protocol utilizes WBC harvested by cytapheresis and then separated into a B, T, NK, and monocyte mixture by H–F gradient separation. In contrast, the AKM protocol further processes the lymphocyte and monocyte mixture obtained from H–F gradients and purifies the monocytes from the lymphocytes by CCE, producing 95% pure monocytes. There is a substantial amount of ongoing research to develop technologies whereby purified effector cell subsets can be rapidly isolated in large numbers directly from the cancer patient. This scientific research approach is summarized in Chapter 11.

Once obtained, purified leukocyte subsets (or mixtures thereof) are then added to culture medium that contains a stimulatory BRM (as reviewed in Table 2); this mixture of leukocyte subset cells and BRM is maintained in a culture vessel for 1 to 30 days. As reviewed in Chapter 12, this period of in vitro manipulation will promote the development of WBC with enhanced tumoricidal capabilities; their numbers may also be expanded as well. After the required incubation time, the cultured cytotoxic cells are removed from the culture flask in preparation for reinjection back into the patient. The AKM, LAK, TIL, or TDAC cells can be injected either intravenously (systemic ACI) or intraperitoneally, intra-arterially, or intrathecally (local/regional ACI), depending on the site of metastatic involvement of the patient; usually, additional maintenance BRM is provided to maintain the cytotoxic capability of the infused cells. In addition, various chemotherapeutic agents may be added to the regimen to ameliorate side effects, provide additive cytotoxic effects, or eliminate suppressor immune system influences. Access to the peritoneal space (for intra-abdominal local/regional ACI) is generally maintained through a Tenckhoff catheter. This system is similar to peritoneal dialysis equipment and is easily maintained by the patient and the nursing staff; other catheter systems exist for other forms of local/regional ACI. Alternatively, intravenous therapy (systemic ACI) requires a permanent indwelling venous catheter device (such as a Hickman catheter) that can be employed for activated cell administration when desired. This indwelling venous catheter can also be used for the apheresis phase of ACI therapy when necessary.

Strategies for the development and testing of ACI therapies have been an area of substantial controversy and confusion since their inception in 1983 (12,13). In general terms, however, Figure 4 presents an overall format for the development of ACI research trials. The ACI trials, to date, have been formulated on basic re-

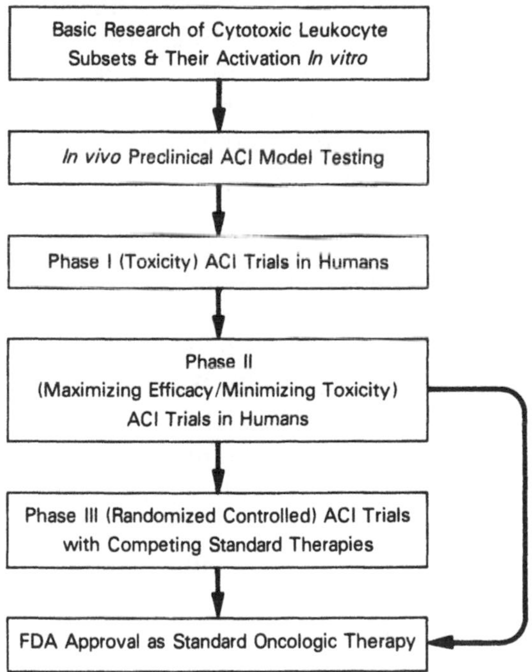

Figure 4 Flowsheet representation of the development of ACI research studies from preclinical observations to potential employment as standard cancer biotherapies.

search findings indicating the cytotoxic potential of specific leukocyte subsets and on studies that have defined the mechanisms by which this cytotoxic capability can be upregulated; principally by the use of BRMs. After the nature of the cytotoxic leukocyte subset–tumor cell interactive process has been studied in vitro, attempts to develop in vivo preclinical models are usually made. These animal models tend to employ either highly inbred strains of mice (thus offering the capacity of transferring immune system cells from one animal to another in an adoptive transfer-type setting) (14) or immunologically impaired nude mice that can receive human tumor xenografts and human immunologic effector cells in an in vivo setting (the nude mouse is effectively an "innocent bystander" while human leukocytes attempt to combat human tumor cells; 15). If in vivo preclinical modeling appears promising, the proposed ACI therapy is then further refined to comply with Food and Drug Administration (FDA) guidelines for the use of activated leukocytes in human cancer biotherapy trials. If these FDA criteria are met and a rational protocol design is passed by local institutional review boards (IRB), Phase I testing of the proposed ACI can begin.

The principal purpose of Phase I testing is to monitor the toxicity of each new therapy in humans by giving escalating doses of the therapy until no longer tolerated. Other Phase I protocol designs have been employed, but it is important to emphasize that the general objective of Phase I testing is to verify the toxicity and safety of the proposed therapy in humans. Phase II ACI trials tend to focus on specific malignancies that (given the preclinical or Phase I information) are believed to be most likely responsive to the proposed ACI therapy. Careful attention is given to administering doses of the ACI therapy below the maximum tolerated dose (MTD) defined in the Phase I trials. The goal of Phase II is to determine the efficacy of the ACI therapy in specific malignancy settings; every attempt to maximize the efficacy and minimize the toxicity of the therapy is made. If the results of Phase II testing indicate that the proposed ACI therapy is clearly superior to any standard therapy available for the malignancy, the FDA will usually grant direct approval for utilization of the new ACI therapy as standard oncologic therapy. However, if the rates of efficacy are not clearly superior to standard therapy (or toxicity is too problematic), a Phase III ACI trial may be indicated to determine the benefit/risk ratio of the proposed ACI therapy compared with standard therapy for the indicated malignancy.

In the subsequent chapters, we will focus on the preclinical and clinical ACI information that has grown out of our understanding of the antitumor activities of NK lymphocytes, blood monocytes and, presumably, cytotoxic T lymphocytes; LAK (chaps. 2-6), AKM (chaps. 7 and 8), and TIL/TDAC (chaps. 9 and 10) immunotherapies, respectively. Also reviewed are strategies for harnessing the maximal creativity of the laboratory and clinical support teams that are crucial to the development of the next generation of ACI therapies (Chaps. 11-13).

REFERENCES

1. Cheaver M A, Greenberg P D, Fefer A. Potential for specific cancer therapy with immune T lymphocytes. J Biol Response Mod 1984; 3:113-127.
2. Ray P K. Suppressor control as a modality of cancer treatment: Perspectives and prospects in the immunotherapy of malignant disease. Plasma Ther Transfusion Technol 1980; 3:101-121.
3. Ray P K, Raychard-Luri S. Immunotherapy of cancer: Present status and future trends. In Ray P K (ed): Immunobiology of Transplantation, Cancer and Pregnancy. New York, Pergamon, 1982:183-207.
4. Oldham R K. Biologicals and biological response modifiers: The fourth modality of cancer treatment. Cancer Treat Rep 1984; 68:221-232.
5. Oldham R K, Smalley R V. Immunotherapy: The old and the new. J Biol Response Mod 1984; 2:1-37.
6. Fefer A. Immunotherapy of cancer. J Natl Cancer Inst 1971; 47:171-189.
7. Fer M F, Beman J, Stevenson H C, et al. A trial of autologous plasma perfusion over protein A in patients with breast cancer. J Biol Response Mod 1984; 3:352-358.

8. Foon K A, Maluish A E, Abrams P G, et al. Recombinant leukocyte α-interferon therapy for advanced hairy cell leukemia: Therapeutic and immunologic results. Am J Med 1984; 80: 351–360.
9. Foon K A, Sherwin S A, Abrams P G, et al. Recombinant leukocyte α-interferon: An effective agent for the treatment of advanced non-Hodgkin's lymphoma. N Engl J Med 1984; 311:1148–1152.
10. Oldham R K. Natural killer cells: History and significance. J Biol Response Mod 1982; 1:217–231.
11. Timonen T, Ortaldo J R, Herberman R B. Characteristics of human large granular lymphocytes and relationship to natural killer and T cells. J Exp Med 1981; 153:569–582.
12. Rosenberg S A, Lotze M T, Muul L M, et al. Observations on the systemic administration of autologous lymphokine-activated killer cells and recombinant interleukin-2 to patients with cancer. N Engl J Med 1985; 313:1485–1492.
13. Stevenson H C, Foon K A, Sugarbaker P. Ex vivo activated monocytes in adoptive immunotherapy trials in colon cancer patients. Prog Clin Biol Res 1986; 211:75–82.
14. Rosenberg S A, Spiess P, La Frenieve R. A new approach to the adoptive immunotherapy of cancer with tumor-infiltrating lymphocytes. Science 1986; 233:1318–1321.
15. Ortaldo J R, Porter H R, Miller P, et al. Adoptive cellular immunotherapy of human ovarian carcinoma xenografts in nude mice. Cancer Res 1986; 46:4414–4419.

2

Lymphokine–Activated Killer Lymphocytes: Evidence for Regulation of Induction and Function by Multiple Cytokines

LAURIE B. OWEN-SCHAUB and ELIZABETH A. GRIMM
M. D. Anderson Hospital, The University of Texas System Cancer Center, Houston, Texas

Several discoveries and technological advances have revolutionized the field of immunology during the past decade. In particular, the rapid progress in recombinant DNA technology has advanced our comprehension of the complex and extensive interactions involving soluble factors in the intact cellular immune response. The availability of recombinant cytokines (soluble factors that transmit growth and differentiation signals between cells) has enabled us to begin dissection of the molecular nature of one of the most intriguing areas of cellular immunology–immunoregulation.

To date, perhaps no single cytokine has proved to be as contributory as interleukin-2 (IL-2) in facilitating our understanding of immunoregulation by soluble factors. Although the functional role of IL-2 was first realized in the proliferation of antigen-specific cytotoxic T lymphocytes (1), it was later shown to be of tantamount importance in the direct activation of "nonclassical" lymphocyte cytotoxicity, termed lymphokine-activated killing (LAK; 2,3). The relative ease of generation, coupled with the fact that LAK is independent of specific antigen recognition, gave impetus to the rapid application of IL-2–activated cells in cancer immunotherapy. Although LAK have been utilized clinically for 5 years, little is known about the regulation of this IL-2–driven cytolytic activity. This chapter will focus on aspects of our current research on IL-2 and the immunoregulatory properties by two other well-characterized cytokines, tumor growth factor-β (TGF-β) and tumor necrosis factor (TNF), as they affect the IL-2–driven induction and maintenance of human LAK oncolysis.

CLINICAL CONSIDERATION OF HUMAN LAK

It is well established that the cellular effector arm of the immune system is primarily responsible for rejection of tumor tissue (4,5). A major hinderance to the development of a successful adoptive immunotherapeutic regimen for the treatment of established human malignancy, however, has been the inability to generate large numbers of cytotoxic lymphocytes with antitumor reactivity. The finding that human lymphocytes could respond directly to IL-2 stimulation with the generation of potent major histocompatibility complex (MHC)-unrestricted cytotoxicity against both autologous and allogeneic tumors provided a favorable prospect for surmounting this impediment (2,3). Largely, on the basis of information obtained from established, successful animal models using LAK cells and recombinant IL-2, clinical trials were initiated. Although several objective tumor responses have been observed (6,7), it is clear that the present treatment modality is far from being optimal and is limited by overt toxic side effects. Many investigators believe that this toxicity is the result of cytokine release from LAK cells or helper T cells that have been stimulated with IL-2. Therefore, it is of paramount importance to elucidate the mechanism of IL-2 activation, the repertoire of cytokines produced after IL-2 stimulation, and their regulatory effect(s) on LAK function.

INTERLEUKIN-2 ACTIVATION OF LAK: LAK EFFECTOR PHENOTYPE

It is well documented that exogenous IL-2 is required for the differentiation of tumor-directed cytotoxic function in peripheral blood lymphocytes (PBL), although correlation of responsiveness with expression of the IL-2 receptor—the Tac antigen—has consistently proved negative (3,8). Recently, our laboratory has provided evidence that a unique IL-2 receptor is involved in the functional differentiation of LAK activity, thus resolving this apparent enigma (9). Chemical cross-linking of ^{125}I-labeled IL-2 to unstimulated PBL followed by sodium dodecyl sulfate-polyacrylamide gel electrophoresis (SDS-PAGE) revealed the presence of a 75-kD, IL-2-binding protein expressed in the absence of the Tac antigen. Cells expressing this alternate IL-2 receptor could transduce the signal for differentiation of NK-resistant oncolysis—called lytic competence—independent of Tac antigen expression. The Tac antigen, although not required for the initial pathway of signal transduction leading to acquisition of function, is rapidly upregulated after IL-2 exposure, and appears to play a role in the proliferative maintenance of LAK cell function. The discrimination of dissimilar IL-2 receptors and developmental stages in LAK generation, namely the acquisition of lytic competence and subsequent proliferative expansion, have proved to be useful for the study of LAK regulation.

Peripheral blood cell-derived LAK function is predominantly mediated by effector cells derived from Leu11 (CD16)-positive lymphocytes [natural killer (NK) lymphocytes]; however, under appropriate culture conditions numerous other lymphocyte subpopulations including B cells, T cells, and "null" cells can express LAK activity (10–12). The unifying feature underlying IL-2 responsiveness by these diverse phenotypic subsets may be the presence of the p75 IL-2 receptor.

To date, no complete study of the contribution of each lymphocyte subpopulation has been reported. Our own investigations strongly suggest that the microenvironmental conditions employed for in vitro LAK generation (particularly the addition of heterologous sera or exogenous cytokines other than IL-2) can greatly influence both the phenotype and function of the effector cell generated. Clearly, much research will be required to understand the regulatory effects of various cytokines on individual lymphocyte subpopulations.

REGULATION OF LAK BY SOLUBLE FACTORS: TRANSFORMING GROWTH FACTOR-β

Although exogenously added IL-2 alone is sufficient for LAK activation, we have observed that other endogenously produced or exogenously added cytokines may participate in suppression and amplification of this effector function. For example, we have observed that transforming growth factor-β (TGF-β) is a potent inhibor of both cytotoxicity and proliferation induced by IL-2 (13). On the other hand, we have reported that interferon-γ addition to IL-2–stimulated cultures can significantly augment LAK effector function (14). Curiously, however, this cytotoxic augmentation was observed to occur only in the presence of monocytes. This finding prompted us to examine the effects of other known interferon-inducible monocyte products on LAK generation and, subsequently, led to the finding that tumor necrosis factor (TNF) elicited potent cytotoxic augmentation. The suppressive and synergistic effects of TGF-β and TNF, respectively, will be discussed in detail in the following sections.

The mechanism(s) by which malignant cells avoid destruction by the immune system of the host are ill-defined. One contributing factor may be existent host-elicited immunosuppression. It has been proposed that progressive growth of any immunogenic tumor beyond a certain critical size will evoke such a state (15). The underlying nature of this immunosuppressive state is highly controversial, but it is thought to involve various soluble suppressor factors. In the LAK system, it is known that serum from cancer patients (16), platelets (13), tumor cells (17), and tumor-derived secretory products (18) can inhibit cytotoxic function. Transforming growth factor-β is a polypeptide with commonality to each of these inhibitory agents.

Transforming growth factor-β, which exerts multiple and diverse biological effects, is a product of both normal and neoplastic cells (for a review see Ref. 19). This cytokine has been shown to stimulate cell growth and differentiation, as well as cause transformation in normal nonlymphoid cells (20,21). On lymphoid cells, TGF-β has been shown to induce potent immune suppression including inhibition of the T-cell mitogenic response (22), interferon-α boosting of NK activity (23), and B-cell immunoglobulin secretion (24). Recently, our laboratory has shown that TGF-β can abrogate IL-2–induced cytotoxicity and proliferation in human LAK cells.

When normal PBL were cultured with IL-2 and TGF-β for 4 days, LAK activity was shown to be decreased in a TGF-β concentration-dependent manner (Table 1). In addition to the diminution of effector cytotoxic function, cellular proliferation was reduced approximately twofold when TGF-β was present (data not shown). Inhibited cell populations that were washed and recultured with IL-2 in the absence of TGF-β expressed LAK activity equivalent to that of controls, indicating that under in vitro conditions the suppressive effects of TGF-β were reversible. No inhibitory effects were observed when TGF-β was added to functional LAK before assay, suggesting that the action of this cytokine occurs during IL-2–induced activation. Interestingly, the capacity of TGF-β to inhibit LAK was directly related to the IL-2 concentration used for activation (Fig. 1). Thus, potent TGF-β–mediated suppression may be expected to occur under conditions of limited IL-2 availability.

Table 1 Transforming Growth Factor-Beta Inhibits LAK Activation

TGF-Beta (ng/ml)	Lytic units/10^6 effectors	% Inhibition
40	1.0	88
20	1.0	88
10	1.2	86
5	3.9	55
1	7.1	17
none	8.6	–

PBL were activated for 5 days in serum-free medium with 100 units/ml IL-2 and the TGF-beta concentrations indicated. Lytic units against the Raji target were calculated as 30% lysis per 10^6 effectors in a 4-hr chromium release assay.

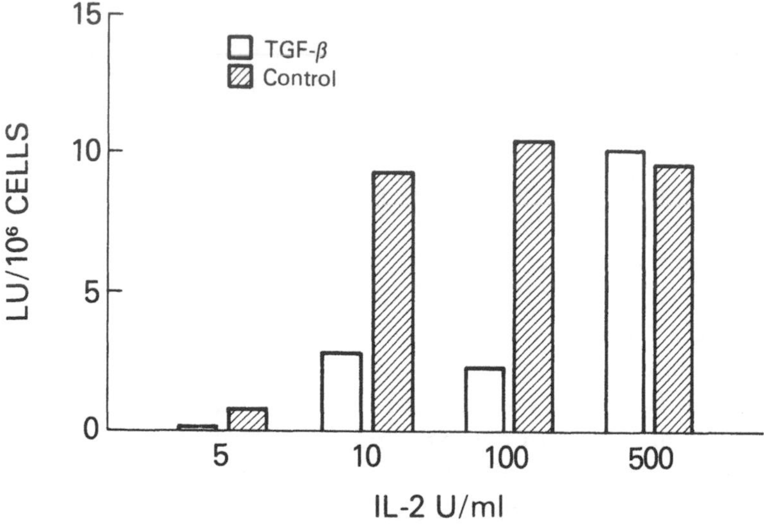

Figure 1 TGF-β suppression of LAK activation is dependent on IL-2 concentration. PBL were cultured in serum-free medium for 5 days with the indicated concentration of IL-2 and TGF-β at a final concentration of 5 ng/ml. Lytic units (LU), as a measure of cytotoxic potency, were calculated per 10^6 effector cells against the Raji target cells in a standard 4-hr chromium release assay. Legend: TGF-β, IL-2 at the indicated concentration + TGF-β at 5 ng/ml. Control, IL-2 alone. TGF-β–mediated potent lytic suppression when IL-2 concentration of <200 units/ml were used. Increasing the IL-2 concentration beyond 200 units/ml rapidly overcame the suppressive effects.

The molecular nature of the TGF-β–induced suppression of LAK activation has not been resolved. In the T-cell system, TGF-β has been shown to prevent the upregulation of both the transferrin receptor and the IL-2 receptor Tac in response to IL-2 stimulation (22). Because these receptors are required for progression through the cell cycle, T-cell proliferation does not occur. As discussed earlier, our laboratory has described two major steps that are required for LAK generation, the first involves the development of lytic competence and occurs independently of proliferation and, the second, requires expression of Tac, followed by active proliferation. If TGF-β interferes with initial signal transduction by the

IL-2 ligand through the 75-kD receptor, upregulation of Tac, as well as the proliferative response would be prohibited. Alternately, TGF-β inhibition may result from the suppression of secondary cytokine production necessary for LAK development. A recent report demonstrated that TGF-β, perhaps in a manner similar to the effect of cyclosporine A, prevented the secretion of interferon-γ, TNF-α, and TNF-β in PBL that had been stimulated with the mitogen phytohemagglutinin (PHA; 25). We have observed positive regulation of LAK effector function in response to exogenous addition of these cytokines, as well as their endogenous production in IL-2-stimulated cultures. These results suggest that TGF-β inhibition of secondary cytokine production may also be operative in the LAK system.

Because of the potential usefulness of LAK or IL-2 in systemic adoptive immunotherapy, it is crucial to understand immune regulation in this system, especially as it applies to the tumor-bearing host. For example, in cancer patients treated with IL-2, only 20% showed an objective clinical response (7), yet all patients' cells could respond to IL-2 stimulation in vitro with potent LAK activity (6). These results imply that downregulation of LAK generation or of effector function may occur in vivo. It has been shown previously that patients' serum, a known source of TGF-β (26), can suppress in vitro LAK activation (16); perhaps TGF-β is responsible for the failure to respond to IL-2 both in vitro and, systemically, in vivo. Additional research is necessary to fully elucidate the mechanisms operative in this apparent anergic response.

In addition to the widespread systemic use of LAK or IL-2, several investigators are employing local administration as a treatment modality. Transforming growth factor-β may similarly regulate LAK in this setting. We have recently described the intracerebral application of LAK and IL-2 in malignant glioma (27,28). Glioblastoma cells are known to secrete a factor closely related to, if not identical with, TGF-β (29,30), and our own results have demonstrated that LAK activation is suppressed in the presence of glioma cells in vitro (17). Although TGF-β would not be expected to inhibit the activated LAK delivered to the brain tumor site in our model, additional recruitment or activation by IL-2 would likely be impaired. Delivery of high-dose IL-2 may circumvent this impairment and allow LAK generation in situ. Understanding the potential limitations inherent in various treatment models will encourage strategies to maximize therapeutic efficacies.

REGULATION OF LAK BY SOLUBLE FACTORS: TUMOR NECROSIS FACTOR

Tumor necrosis factor (TNF) was first described as an endotoxin-induced serum mediator causing hemorrhagic necrosis of murine tumors in vivo (31). The biologically active substance responsible for this reaction was shown to be a protein product derived from activated macrophages (32) and lymphocytes (33) and was named TNF-α. This cytokine is one member of a family of proteins known to be cytotoxic/cytostatic for neoplastic cells. Tumor necrosis factor-β (TNF-β), also known as lymphotoxin, is another such lymphocyte protein product sharing similar functions (34, 35) and amino acid homology (34,36) to TNF-α. In additional to the direct tumoricidal activities of TNF (35,37), this cytokine family has been shown to stimulate fibroblast growth (38), enhance surface antigen expression on tumor (39) and endothelial cells (40), activate neutrophils (41), augment proliferation of activated T cells (42), and enhance macrophage (43) and NK (44) cytotoxicity. Our recent finding that exogenous interferon-γ can augment LAK function (17), taken together with the observation that IL-2–induced TNF production and its further enhanced by interferon-γ (45), led to our examination of the effects of TNF on LAK activity.

Optimal LAK activity is routinely observed in vitro at IL-2 concentrations of > 100 units/ml (2.2 nmol). Use of IL-2 concentrations as low as 10 units/ml produces significant, yet clearly suboptimal, lysis of NK-resistant targets. When PBL were cultured with suboptimal IL-2 concentrations and TNF, a marked augmentation of LAK activity was observed (46). This augmentation was seen with both TNF-α and TNF-β and was maximal when TNF was used at approximately 500 units/ml. Culture of PBL with TNF, in the absence of IL-2, did not generate LAK activity. Thus, the augmentation observed with IL-2–TNF was a true synergistic response. Of particular note, this synergistic enhancement enabled the generation of optimal LAK function (equal to or exceeding that with 100 units/ml of IL-2 alone) at a tenfold lower IL-2 concentration (Fig. 2). Synergy was detectable with both fresh tumor preparations of various histological types, as well as cultured tumor targets. The synergistic effects of IL-2–TNF occurred over a range of IL-2 concentrations, but they were most notable when lower IL-2 concentrations were used for activation (Table 2). For most PBL tested, no further enhancement was seen when optimal IL-2 concentrations (> 100 units/ml) were employed.

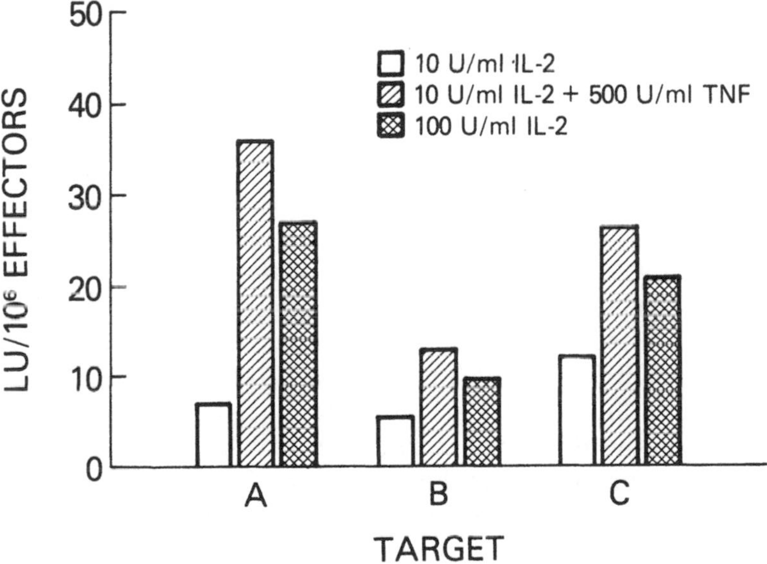

Figure 2 TNF-α augmentation of IL-2-activated PBL cytotoxicity against fresh and cultured tumor targets. PBL were activated in serum-free medium for 5 days with the indicated concentrations of IL-2 and TNF-α. Lytic units (LU), as an indication of cytotoxic potency, were calculated per 10^6 effector cells in a 4-hr chromium release assay. Tumor targets shown include (A) fresh adenocarcinoma preparation, (B) fresh sarcoma preparation, and (C) cultured Burkitt lymphoma cell line, Raji. In every case, 10 units IL-2 + TNF resulted in greater lytic activity than 100 units IL-2 alone.

Table 2 TNF-α Synergy Is IL-2 Dose-Dependent

IL-2 concentration (U/ml)	Lytic units/10^6 effectors	
	-TNF	+TNF
0	<1	<1
10	14	39
100	24	44
500	49	48

PBL were activated for 5 days in serum-free medium at the indicated IL-2 concentration with or without TNF-α at 500 units/ml. Lytic units against the Raji cells were calculated as 30% lysis per 10^6 effectors in a standard chromium release assay.

In the T-cell system, TNF has been shown to augment antigen-stimulated proliferation, presumably by increasing high-affinity IL-2 receptors (42). In the LAK system, however, no detectable increase in thymidine incorporation or cell recovery was apparent when PBL that were cultured with IL-2–TNF were compared with those cultured with IL-2 alone (46). Despite the fact that no augmentation of cellular proliferation was detectable, IL-2–TNF stimulation resulted in an increased percentage of IL-2 receptor-bearing cells identified by the Tac epitope. It is conceivable that upregulation of this receptor may facilitate heightened IL-2 responsiveness in the absence of increased cell cycling. Alternately, it is possible that a minor lymphocyte subpopulation with potent effector function is undergoing proliferative enhancement and that this effect is not observable at the population level.

Biological responsiveness to TNF is thought to require interaction with specific cell membrane receptors (35,37). Our laboratory has demonstrated the de novo expression of TNF receptors on PBL after IL-2 stimulation (47). Induction of this receptor occurs in both a time- and an IL-2 concentration-dependent manner, with maximal expression observable at 5 days and 1000 units/ml IL-2. The distribution of this receptor is not limited to a particular phenotypic subset, with NK, T-helper, and T-cytotoxic cells uniformly demonstrating induction. Biological responsiveness has not yet been directly correlated with receptor expression in these various lymphocyte subsets. This is currently an area of active investigation in our laboratory.

The ability of IL-2 to induce TNF receptors on PBL led us to examine TNF production in these cultures. Our results clearly demonstrate an IL-2 dose-dependent production of both TNF-α and TNF-β by PBL. Inclusion of neutralizing antibodies against these TNF species in IL-2-stimulated PBL cultures significantly reduced the generation of LAK. Because the presence of TNF in the culture milieu contributes to LAK generation, as well as enhances IL-2-induced cytotoxicity under appropriate conditions, we have proposed that TNF may be part of a positive-feedback circuit operative in the antigen-independent activation and regulation of human cytotoxic lymphocytes.

CONCLUSIONS

Although adoptive immunotherapy with LAK cells and recombinant IL-2 can mediate the regression of various tumor types (6,7), this treatment modality is associated with substantial toxicity. In the current regimen, the major side effects result from the administration of high-dose IL-2. The IL-2–TNF synergy described permits the production and maintenance of optimal LAK activity in vitro with a tenfold reduction in IL-2. If, in vivo, the function of IL-2–TNF is synergistic, then decreased toxic side effects of IL-2 resulting in improved therapeutic efficacy may be possible.

Because TNF is produced by PBL after IL-2 stimulation, and it has been shown to cause unwanted side effects at high doses as a single therapy (48), it is possible that TNF production is directly responsible for a portion of the toxicity observed after high-dose IL-2 therapy. Hence, it will be imperative to monitor endogenous TNF production in response to IL-2 administration when attempting combination cytokine therapy. However, with proper patient monitoring and dose scheduling, combination cytokine administration may be a viable alternative to existing modalities.

The documented inhibitory activity of TGF-β and the synergistic action of TNF on LAK activity highlight the need for a greater understanding of the scores of other understudied cytokines on this unique cytotoxic function. As an outgrowth of such research, we may be able to develop effective strategies by which "mixed signals" will be avoided, thus maximizing the LAK activity generated in vitro and maintained in vivo. Similarly, each of these regulatory cytokines have potential individual toxic effects that may be additive to one another: A firm understanding of the cytokine-mediated side effects that must be tolerated for therapeutic efficacy (by virtue of the essential need for a cytokine in induction or maintenance of LAK activity), and those that must be actively eliminated or suppressed, will help to minimize the risks and toxicities of LAK therapies in the future.

REFERENCES

1. Baker P E, Gillis S, Ferm M, Smith K A. The effect of T-cell growth factor on the generation of cytolytic T-cells. J Immunol 1978; 121:2168.
2. Grimm E A, Mazumder A, Zhang H Z, Rosenberg S A. Lymphokine-activated killer cell phenomenon. Lysis of natural killer resistant fresh solid tumor cells by interleukin-2 activated autologous human peripheral blood lymphocytes. J Exp Med 1982; 155:1823.
3. Grimm E A, Rosenberg, S A. The human lymphokine-activated killer cell phenomenon. In Pick E (ed): *Lymphokines*. New York, Academic Press, 1984:279–311.
4. Hellstrom I, Hellstrom K E. Studies on cellular immunity and its serum-mediated inhibition in moloney virus-induced mouse sarcomas. Int J Cancer 1969; 4:587.
5. Cerottini J C, Brunner K T. Cell-mediated cytotoxicity, allograft rejection, and tumor immunity. Adv Immunol 1974; 18:67.
6. Rosenberg S A, Lotze M T, Muul L M, Leitman S, Chang A E, Ettinghausen S E, Matory Y L, Skibber J M, Shiloni E, Vetto J T, Seipp C A, Simpson C, Reichert C M. Observations on the

systemic administration of autologous lymphokine activated killer cells in recombinant IL-2 to patients with metastatic cancer. N Engl J Med 1985; 313:1485-1492.

7. Lotze M T, Chang A E, Seipp C A, Simpson C, Vetto J T, Rosenberg S A. High dose recombinant interleukin-2 in the treatment of patients with disseminated cancer: Responses, treatment related morbidity, and histologic findings. JAMA 1986; 256:3117.
8. Gray J D, Shau H, Golub S H. Functional studies on the precursors of human lymphokine-activated killer cells. Cell Immunol 1985; 96:338.
9. Owen-Schaub L B, Loudon W G, Yagita M, Grimm E A. Functional differentiation of human lymphokine activated killing (LAK) is distinct from expansion and involves dissimilar interleukin-2 receptors. Cell Immunol 1988; 111:235.
10. Damle N K, Doyle L V, Bradley E C. Interleukin-2 activated human killer cells are derived from phenotypically heterogenous precursors. J Immunol 1986; 137:2814.
11. Ortaldo J R, Mason A, Overton R. Lymphokine-activated killer cells. Analysis of progenitors and effectors. J Exp Med 1986; 164:1193.
12. Phillips J H, Lanier L L. Dissection of the lymphokine-activated killer phenomenon. Relative contribution of peripheral blood natural killer cells and T lymphocytes to cytolysis. J Exp Med 1986; 164:814.
13. Grimm E A, Ammann R, Crump W L III, Durett A, Hester J P, Lagoo-Deenadalayan S, Owen-Schaub L B. TGF-beta inhibits the in vitro induction of lymphokine activated killing activity. Cancer Immunol Immunother 1988; 27:53.
14. Grimm E A, Owen-Schaub L B, Loudon W G, Yagita M. Lymphokine-activated killer cells: Induction and function. Ann NY Acad Sci 1987; 532:380.
15. North R J. Down-regulation of the anti-tumor immune response. In Klein G, Weinhouse S (eds): Advances in Cancer Research. New York, Academic Press, 1985:1.
16. Itoh K, Tilden A B, Balch C M. Role of interleukin-2 and a serum suppressive factor in the induction of activated killer

cells cytotoxic for autologous human melanoma cells. Cancer Res 1985;45:3173.

17. Grimm E A. Human lymphokine-activated killer cells (LAK cells) as a potential immunotherapeutic modality. Biochim Biophys Acta 1987;865:26.
18. Roth J A, Grimm E A, Gupta R K, Ames R S. Immunoregulatory factors derived from human tumors: I. Immunologic and biochemical characterization of a suppresses lymphocyte proliferative and cytotoxic in vitro. J Immunol 1982; 128:1975.
19. Sporn M B, Roberts A B, Wakefield L M, Assoian R K. Transforming growth factor-beta: Biological function and chemical structure. Science 1986;233:532.
20. Anzano M A, Roberts A B, Smith J M, Sporn M B, DeLarco J E. Sarcoma growth factor from conditioned medium of virally transformed cells is composed of both type α and type β transforming growth factors. Proc Natl Acad Sci USA 1983;80:6264.
21. Anzano M A, Roberts A B, DeLarco J E, Wakefield L M, Assoian R K, Roche N S, Smith J M, Lazarus J E, Sporn M B. Increased secretion of type beta transforming growth factor accompanies viral transformation of cells. Mol Cell Biol 1985; 5:242.
22. Kehrl J H, Wakefield L M, Roberts A B, Jakoelew S, Alvarez-Mon A, Derynck R, Sporn M B, Fausi A S. Production of transforming growth factor-beta by human T lymphocytes and its potential role in the regulation of T cell growth. J Exp Med 1986;163:1037.
23. Rook A H, Kehrl J H, Wakefield L M, Roberts A B, Sporn M B, Burlington D B, Lane H C, Fauci A S. Effects of transforming growth factor-beta on the function of natural killer cells: Depressed cytolytic activity and blunting of interferon responsiveness. J Immunol 1986;136:3916.
24. Kehrl J H, Roberts A B, Wakefield L M, Jakowlew S, Sporn M B, Fauci A S. Transforming growth factor-beta is an important immunomodulatory protein for human B lymphocytes. J Immunol 1986;137:3855.

25. Espevik T, Figari I S, Shalaby M R, Lackides G A, Lewis A D, Shepard H M, Palladino M A. Inhibition of cytokine production by cyclosporine A and transforming growth factor-beta. J Exp Med 1987; 166:571.
26. Childs C B, Proper J A, Tucker R F, Moses H L. Serum contains a platelet-derived transforming growth factor. Proc Natl Acad Sci USA 1982; 79:5312.
27. Jacobs S K, Wilson D J, Kornblith P L, Grimm E A. Interleukin-2 and autologous lymphokine-activated killer cells in the treatment of malignant glioma. J Neurosurg 1986; 64: 743.
28. Jacobs S K, Wilson D J, Kornblith P L, Grimm E A. Interleukin-2 or autologous lymphokine-activated killer cell treatment of malignant glioma: Phase I trial. Cancer Res 1986; 46:2101.
29. Kikuchi K, Neuwelt E A. Presence of immunosuppressive factors in brain-tumor cyst fluid. J Neurosurg 1983; 59:790.
30. Wrann M, Bodmer S, de Martin R, Siepl C, Hofer-Warbinek R, Frie K, Hofer E, and Fontana A. T cell suppressor factor from human glioblastoma cells is a 12.5 kD protein closely related to transforming growth factor-beta. EMBO J 1987; 6:1633.
31. Carswell E A, Old L J, Kassel R L, Green S, Fiore N, Williamson B. An endotoxin induced serum factor that causes necrosis of tumors. Proc Natl Acad Sci USA 1975; 25:3666.
32. Mannel D M, Moore R N, Mergenhagen S E. Macrophages as a source of tumoricidal activity (tumor necrotizing factor). Infect Immun 1980; 30:523.
33. Peters P M, Ortaldo J R, Shalaby M R, Svedersky L P, Nedwin G E, Bringman T S, Hass P E, Aggarwal B B, Herberman R B, Goeddel D V, Palladino M A. Natural killer-sensitive targets stimulate production of TNF-alpha but not TNF-beta (lymphotoxin) by highly purified peripheral blood large granular lymphocytes. J Immunol 1986; 137:2592.
34. Gray P W, Aggarwal B B, Benton C V, Bringman T S, Henzel W J, Jarrett J A, Leung D W, Moffat B, Ng P, Svedersky L P, Palladino M A, Nedwin G E. Cloning and expression of cDNA

for human lymphotoxin, a lymphokine with tumor necrosis activity. Nature 1984; 321:721.

35. Sugarman B J, Aggarwal B B, Hass P E, Figari I S, Palladino M A, Shepard H M. Recombinant human necrosis factor-alpha: Effects on proliferation of normal and transformed cells in vitro. Science 1985; 230:943.
36. Pennica D, Nedwin G E, Hayflick J S, Seeberg P H, Derynck R, Palladino M A, Kohr W J, Aggarwal B B, Nedwin G E. Human tumor necrosis factor: Precursor structure, expression, and homology to lymphotoxin. Nature 1984; 312:721.
37. Ruggiero V, Latham K, Baglioni C. Cytostatic and cytotoxic activity of tumor necrosis factor on human cancer cells. J Immunol 1987; 138:2711.
38. Vilcek J, Palombella V J, Hendriksen-DeStefano D, Swenson C, Feinamn R, Hirai M, Tsujimoto M. Fibroblast growth enhancing activity of tumor necrosis factor and its relationship to other polypeptide growth factors. J Exp Med 1986; 163: 632.
39. Scheurich P, Kronke M, Schulter C, Ucer U, Pfizenmaier K. Noncytocidal mechanisms of action of tumor necrosis factor-alpha on human tumor cells: Enhancement of HLA gene expression synergistic with interferon gamma. Immunobiology 1986; 172:291.
40. Porber J S, Bevilacqua M P, Mendrick D L, Lapierre L A, Fiers W, Gimbrone M A Jr. Two distinct monokines, interleukin 1 and tumor necrosis factor, each independently induce biosynthesis and transient expression of the same antigen on the surface of cultured human vascular endothelial cells. J Immunol 1986; 136:1680.
41. Shalaby M, Aggarwal B B, Rinderknecht E, Svedersky L P, Finkle B S, Palladino M A. Activation of human polymorphonuclear neutrophil functions by interferon-gamma and tumor necrosis factors. J Immunol 1985; 135:2069.
42. Scheurich P, Thomas B, Ucer U, Pfizenmaier K. Immunoregulatory activity of recombinant human tumor necrosis factor (TNF)-alpha: Induction of TNF receptors on human T cells

and TNF-alpha-mediated enhancement of T cell responses. J Immunol 1987; 138:1786.
43. Phillip R, Epstein L B. Tumor necrosis factor as an immunomodulator and mediator of monocyte cytotoxicity induced by itself, gamma interferon, and interleukin 1. Nature 1986; 323:86.
44. Ostensen M E, Thiele D L, Lipsky P E. Tumor necrosis factor-alpha enhances cytolytic activity of human natural killer cells. J Immunol 1987; 138:4185.
45. Nedwin G E, Svedersky L P, Bringman T S, Palladino M A, Goeddel D V. Effect of interleukin 2, interferon-gamma, and mitogens on the production of tumor necrosis factors alpha and beta. J Immunol 1985; 135:1354.
46. Owen-Schaub L B, Gutterman J U, Grimm E A. Synergy of TNF in the activation of human cytotoxic lymphocytes. Tumor necrosis factor is synergistic with interleukin-2 in the generation of human lymphokine activated killer cell cytotoxicity. Cancer Res 1988; 48:788.
47. Owen-Schaub L B, Crump W L III, Morin G I, Grimm E. A. TNF binding to IL-2 stimulated peripheral blood lymphocytes: Analysis by direct binding, chemical cross-linking, immunofluorescence and immunoprecipitation. J Immunol 1989; (submitted).
48. Blick M, Sherwin S A, Rosenblum M, Gutterman J U. Phase I study of recombinant tumor necrosis factor in cancer patients. Cancer Res 1987; 47:2986.

3

Lymphokine-Activated Killer Lymphocytes: National Cancer Institute Clinical Trials

JEFFREY CLARK and DAN L. LONGO
National Cancer Institute, Frederick, Maryland

A number of approaches have been used to enhance the immune response of cancer patients against their tumors in an attempt to treat their malignancies. These have included methods aimed at either inducing an immune response specifically directed against the cancer cells by using tumor cell or tumor antigen vaccines or have used a variety of nonspecific immune stimulants including bacterial products such as bacillus Calmette-Guerin (BCG) or *Corynebacterium parvum*; chemical immunodulators, such as levamisole; interferon inducers such as poly-ICLC; or natural human products with immunomodulatory properties, such as interferons and thymosins, with the hope that through a general stimulation of the immune system a substantial immune response against the tumor could also be achieved. A number of biological agents have been tested against a wide variety of cancers, with only limited success. This is, at least in part, due to the inability to stimulate a sufficiently strong hosts' immune response against their own malignant cells. However, the availability of an ever-increasing number of recombinant compounds, with potent effects on various effector mechanisms of the immune system, and an ever-increasing understanding of the mechanisms involved in immune-mediated antitumor effects, have again raised hope that these agents, or combination of these agents with each other and with other forms of therapy, might be able to induce and maintain a sufficient antitumor immune response to yield decisive clinical responses against tumors.

Another method of enhancing the efficacy of immunotherapy in the treatment of cancer is to directly activate immune cells ex vivo that can then be infused into the patient to mediate an antitumor response, either directly or indirectly. Either autologous or heterologous cells could be used, although autologous cells should be preferable, to decrease any host response that might be directed against the infused cells. The use of an already active form of immunotherapy that is transferred to a patient is called *adoptive cellular immunotherapy* (ACI). A variety of cells in the immune system, for example, T cells, natural killer (NK) cells, lymphokine-activated killer (LAK) cells, monocytes–macrophages, or B cells, could be isolated from patients by leukapheresis, activated, and reinfused into patients in an attempt to directly mediate tumor regression. The activation of cells outside of the body before transferring them to the patient has the advantages of the ability to generate many activated cells and not depend on a completely intact host immune system for efficacy.

An approach of adoptive cellular immunotherapy, initially piloted by Rosenberg at the NCI, utilizes recombinant interleukin-2 (rIL-2) and lymphokine-activated killer cells (LAK cells; 1). Interleukin-2 is a lymphokine produced by T cells with multiple critical effects on the immune system, including activation of other T cells, natural killer (NK) cells, monocytes, and B cells (2, 3). The LAK activity is mediated by blood leukocytes that are lytic for a broad spectrum of fresh autologous, syngeneic, or allogeneic primary or metastatic tumor cells in vitro (4). These cells are also capable of lysing NK-resistant Daudi cells. Although the ultimate determination of the nature of the cells that can be precursors for cells with LAK activity is still ongoing, these cells do not have T- or B-lymphocyte markers and appear to consist of several lymphocyte subsets (5–10). After activation with IL-2, most of the LAK activity appears to be mediated by cells bearing NK surface markers, but a minor contribution also comes from T^{3+},

T^{8+} cells (5–10). Most LAK effector cells bear the Leu19 surface antigen (8).

The in vitro studies indicating the ability of LAK cells to lyse fresh tumor targets led to a number of studies using LAK plus IL-2 to treat cancers in animals. These showed a reduction in the size and number of established pulmonary or hepatic metastases in mice given LAK and IL-2 (11,12). Although a decrease in the number of metastases was also seen when high doses of IL-2 alone were used, at any given dose of IL-2, up to the very highest, a significantly greater reduction in the number of metastases was seen when IL-2 plus LAK cells were used in combination. Subsequent studies in mice have also suggested a correlation between tumor response to IL-2 plus LAK therapy and LAK cell-mediated cytolysis in vitro (13).

SYSTEMIC LAK PLUS INTERLEUKIN-2 THERAPY

The initial trials in humans used IL-2 alone in escalating doses and LAK cells alone in escalating doses to evaluate the toxicity plus efficacy of each of the components of the treatment regimen. No responses were seen in patients who received either IL-2 alone or LAK cells alone (1). However, when IL-2 was subsequently given at high doses for a more prolonged period, clinical responses were seen and, as will be discussed later, one of the open questions about IL-2 plus LAK cell therapy remains on what role the LAK cells, generated ex vivo and then returned to the patient, play in the antitumor effect of treatment, as opposed to the role played by the high doses of IL-2 alone (14). This question can be addressed only by randomized clinical trials, and these are currently ongoing.

The next step in the development of the treatment schedule was the combination of the high doses of IL-2 with LAK cells. Two IL-2 plus LAK treatment regimens have been developed that are currently the major ones being used to treat patients. In the initial regimen developed at the National Cancer Institute (NCI) in pilot studies that use this therapy to treat patients with advanced cancer, a standard cycle of therapy consists of 16 days of treatment (Table 1) (14). On days 1 through 5 the patient receives IL-2 at a

Table 1 LAK + IL-2 Regimens

1. Standard IV bolus IL-2 systemic regimen	
days 1-5:	IL-2 (100,000 U/kg) IV bolus q 8 hr
days 6-7:	no treatment
days 8-12:	leukapheresis 4 hr each day with cells grown in IL-2
day 12 (postpheresis):	begin IL-2 (100,000 U/kg) IV bolus and continue as long as tolerated (mean of nine doses)
day 12:	day 8 and 9, LAK cells along with IL-2
day 13:	day 10, LAK cells
day 15:	day 11, and 12 LAK cells
2. Continuous infusion IL-2 systemic regimen	
days 1-5:	continuous infusion (3×10^6 U/m^2 per day) IV IL-2
day 6:	(36 hr)—no treatment
days 7-10:	leukapheresis with cells grown in IL-2
days 10-15:	continuous infusion (3×10^6 U/m^2 per day) IV IL-2 for 5 days
day 10:	day 7, LAK cells
day 11:	day 8, LAK cells
day 13:	day 9 and 10, LAK cells
3. IP LAK + IL-2	
days 1-3:	IL-2 (100,000 U/kg) IV bolus q 8 hr
days 4-5:	no treatment
days 6-10:	leukapheresis 4 hr each day with cells grown in IL-2
days 11-12:	no treatment
days 13-17:	IL-2 (25,000 U/kg) IP bolus q 8 hr plus LAK cells IP—1 bag each day administered with a.m. IL-2
days 18-19:	no treatment
days 20-24:	leukapheresis 4 hr each day with cells grown in IL-2
days 25-26:	no treatment
days 27-31:	IL-2 (25,000 U/kg) IP bolus q 8 hr plus LAK cells IP—1 bag each day administered with a.m. IL-2

dose of 100,000 units/kg every 8 hr intravenously. During this priming period, the peripheral blood lymphocyte count and the number of circulating LAK precursors decline, but when the IL-2 is stopped, there is a rebound increase in both the total peripheral blood lymphocyte count and the number of LAK cell precursors in the blood above baseline levels, which occurs within 24 hr after discontinuation of IL-2 and persists for several days. On days 6 and 7 the patient receives no treatment. On each day 8 through 12, the patient has a 4-hr leukapheresis, and those peripheral blood mononuclear cells that have rebounded are removed and incubated with IL-2 to activate them. On the 12th day, the last day of leukapheresis, the activated LAK cells that have been generated from the peripheral blood cells obtained on the first and second days of leukapheresis (days 8 and 9 of therapy) are reinfused along with systemic IL-2 at the same dose of 100,000 units/kg intravenously every 8 hr. On the 13th day of treatment, the LAK cells generated from the third day of leukapheresis are reinfused along with systemic IL-2 and, again, on the 15th day, the patient receives the remaining LAK cells that were generated from the last 2 days of leukapheresis. Systemic IL-2 is continued as long as it is tolerated by the patient.

In addition to IL-2 and LAK, the regimen also includes indomethacin and acetaminophen (for suppression of fevers), ranitidine or cimetidine (to prevent stress gastritis), prochlorperazine (Compazine) and other antiemetics as needed, meperidine (Demerol) as needed for the treatment of severe chills, low-dose dopamine for oliguria, and phenylephrine for the treatment of hypotension. Because indomethacin has beneficial effects on LAK cell effectiveness in vitro (most likely through inhibition of prostaglandin effects), ranitidine or cimetidine can block T-suppressor cell function (which has been shown to be important in inhibiting LAK function) in vitro, and prochlorperazine has been shown to enhance LAK function in vitro, it is possible that any or all of these agents may also contribute to the therapeutic effects of IL-2 and LAK therapy (15–18).

The clinical treatment regimen using IL-2 and LAK therapy requires technological and clinical expertise because these cells must be generated in vitro in large numbers and the toxicity associated with this treatment is substantial. Most patients must receive IL-2 and LAK in an intensive care unit (ICU) with continuous cardiac and blood pressure monitoring to monitor and treat the side effects resulting from the treatment. The exact mechanisms of all of the adverse effects seen have not been clearly elucidated. However, one major mechanism appears to be related to increased capillary permeability induced by IL-2 or a lymphokine or cytokine product of IL-2–stimulated cells, leading to fluid shifts into extravascular tissues. Related to this, over 50% of patients develop hypotension that requires vasopressors, prerenal azotemia with anuria (defined as a urine output less than 10 ml/hr), and elevated serum creatinine levels, and over a third of patients gain over 10% of their body weight in retained fluid, with a substantial proportion of other patients gaining < 10% body weight from fluid retention (Table 2; 1,11,19–21). Other adverse effects include fever or chills, nausea or vomiting, and fatigue in most patients; elevated liver function tests (both transaminases and bilirubin); diffuse erythematous skin rash with or without pruritus; anemia requiring red blood cell transfusions; thrombocytopenia; alterations in mental function (ranging from anxiety to confusion to psychosis), and eosinophilia (which can be marked), all occur in approximately 50% or more of patients. At least some of the hematological toxic responses (anemia, thrombocytopenia, and decreased granulocytes) seen with IL-2 plus LAK treatment may be due to suppression of hematopoiesis by IL-2 or IL-2–stimulated lymphokines such as interferon-γ (22). In the one analysis that has been reported, there was no relationship between the percentage of eosinophils in the peripheral blood and the toxic responses, but this is an area that needs further study before firm conclusions can be drawn about the possible role of eosinophils in any of the IL-2–related toxicity (23). A significant percentage of patients complain of mild dyspnea, but serious pulmonary problems are seen in less than 20% of patients with approximately 5% of patients overall having required intubation. Cardiac arrhythmias, primarily atrial, although ventricular arrhythmias are also seen, have occurred in about 13%, and either angina or myocardial infarctions, are

Table 2 Systemic LAK + IL-2 Toxicities

Almost all patients (>90%)
fatigue
fevers/chills
nausea/vomiting
Most patients (>50%)
hypotension
diffuse erythematous skin rash/pruritus
weight gain
diarrhea
nasal congestion
abnormal renal function (oliguria or creatintine level >2.0)
abnormal liver function (hyperbilirubinemia, elevated SGOT or alkaline phosphatase levels)
thrombocytopenia (platelets <100,000)
anemia requiring transfusion
CNS disturbance (anxiety/somnolence/disorientation/inappropriate behavior)
eosinophilia
Uncommon (<20%)
severe respiratory distress
cardiac arrhythmias (primarily atrial, although ventricular also seen)
central line sepsis
severe mucositis
coma
myocardial ischemia
Rare (<5%)
GI hemorrhage
pleural effusion requiring thoracentesis
seizure
myocardial infarction
death (myocardial infarction; pulmonary insufficiency; plus sepsis)

promptly reversible upon discontinuation of the IL-2. The use of corticosteroids (dexamethasone 4 mg every 6 hr) has been shown to decrease certain of the side effects (dyspnea, fever, pruritus, confusion, elevated BUN and creatinine levels) without significantly affecting hematological toxic effects or the weight gain seen in IL-2 plus LAK therapy (24). No antitumor responses were seen in six patients treated, but only three of these had malignancies (melanoma) with known reasonable response rates to IL-2 plus LAK therapy. Although there must be concern that steroids, even though they decrease some of the toxicity, might adversely affect the therapeutic efficacy of IL-2 plus LAK, this question has not yet been completely answered.

Hepatitis A infections were seen in several of the IL-2 plus LAK-treated patients, but this appeared to be related to contamination of a single lot of the medium (which contained human sera) used for growing the LAK cells (14,25). Although this emphasizes the critical need for a careful sterile procedure in the production of and screening (even for viruses not usually associated with transfusion transmission) of any human serum component used in the treatment of patients, this appears to have been a self-limited adverse effect. It did provide a major impetus for research into the use of serum-free media, which has been shown to be sufficient for the effective in vitro growth of LAK cells (26).

Several tumor types have been treated with this regimen. It appears to be most effective in renal cell carcinoma (especially in patients with a low tumor burden), melanoma, non-Hodgkin's lymphoma (NHL) and, to a lesser extent, colon cancer (Table 2; 1,14,27,28). With use of the standard IL-2 plus LAK regimen as just defined, response rates in renal cell carcinoma range from 15% to 33%, with an overall response rate of 25%. Approximately 21% of melanoma patients have responded, and 13% of patients with colorectal carcinoma have responded. Two patients with follicular non-Hodgkin's lymphoma have had a response. The durability of responses has ranged from 1 month to longer than 2 years, but most responses last for 4 to 6 months. Of note, most of the patients who achieved a complete response have maintained that response, to date, with the patient furthest out still in remission at

22 months (11). This therapy given systemically has been tried in a limited number of patients with sarcomas and adenocarcinoma of the lung, and in one patient each with esophageal, ovarian, pancreatic, and gastric cancer. Although minor responses have been seen, no partial or complete responses have been seen in these tumors.

The other major systemic IL-2 plus LAK cell therapy treatment regimen that has been reported used a modification of the initial regimen. Basically, the IL-2 used in priming the patients and in the portion of the regimen combined with the LAK cells was given by continuous infusion, rather than by bolus (29). The dose of IL-2 was escalated from 1×10^6 units/m^2 over 24 hr IV daily for 5 days to 3×10^6 and 5×10^6 units/m^2 as tolerated in the final patients on the study. After the 5 days of continuously infused IL-2 therapy, the patients received no treatment for 36 hr, followed by leukapheresis on days 7 through 10 with cells incubated with IL-2. Cells from day 7 were given back on day 10, cells from day 8 on day 11, and cells for day 9 and 10 given back together on day 13. The IL-2 was again given at similar doses by continuous infusion for 5 days, beginning after leukapheresis on day 10. The therapy was repeated for responders.

Toxic reactions were qualitatively similar to, although quantitatively significantly less than, those seen with the bolus IL-2 plus LAK regimen. Only 6 of 40 patients had to be transferred to the ICU for treatment at some point in their course. No myocardial infarctions were seen, and there was one treatment-related death (2.5%), caused by sepsis.

Responses were seen in five of ten melanoma patients, three of six renal cell carcinoma patients, and in one patient each with lung, parietal, and ovarian cancers and in one patient each with non-Hodgkin's lymphoma and Hodgkin's disease (Table 3). None of the 13 patients with colon cancer responded. All of the responses were partial, and most were of short duration (≤ 4 months). The time from start of therapy to maximum response was short (15–30 days). Dosages $\geq 3 \times 10^6$ units/m^2 daily gave greater lymphocytosis than did the lower dosages, although increasing the dose above 3×10^6 units/m^2 did not significantly

Table 3 Systemic LAK + IL-2

	Results for selected malignancies								
	(1)			(2)			(3)		
	Pr[a]	Cr[b]	Overall	Pr	Cr	Overall	Pr	Cr	Overall
Melanoma	15	8	23	19	0	19	50	0	50
RCC[c]	22	11	33	12	3	15	50	0	50
Colorectal	8	4	12				0	0	0

[a]Pr, partial response (greater than 50% decrease in measurable tumor).
[b]Cr, complete response (no evidence of measurable tumor).
[c]RCC, renal cell carcinoma.
Sources: (1)Rosenberg et al. (Ref. 14), (2) Dutcher et al. (Ref. 28) and Fisher et al. (Ref. 27), (3) West et al. (Ref. 29).

increase rebound lymphocytosis. There was a direct correlation between response and both the baseline lymphocyte count and the absolute rebound lymphocytosis after IL-2 treatment.

It is now too early to know which of these regimens is better. Reasonable response rates are seen with each of these regimens for both melanoma and renal cell cancer patients, and too few patients have been treated with the two regimens, for too short a time, to make any valid analysis of differences in response rates (complete or partial) or duration of responses. The kinetics of IL-2 indicate a very short initial serum half-life (6–7 min), as would be expected for a relatively small protein. If having IL-2 continuously present during the period when LAK cells are reinfused is important for their activity and survival, then the kinetics of IL-2 suggest that continuous infusion of IL-2 would be preferable. However, it is possible that high peak levels of IL-2 are somehow important (possibly to deliver sufficient IL-2 levels to cells directly at tumor sites) in mechanisms of the IL-2 response, which would mean that bolus therapy might be preferable. In addition, kinetic studies on serum IL-2 levels in patients receiving IL-2 therapy do suggest that a high bolus dose IL-2 (100,000 units/kg) given every 8 hr does yield detectable serum IL-2 levels throughout that 8-hr period, although these have dropped to low levels (1–5 units/ml) by the end of the 8 hr. There are ongoing studies addressing these questions. There is preliminary evidence that does suggest that 5 days of priming with IL-2 (whether by the bolus regimen or by continuous infusion, as is done in both current systemic regimens) is better than 3 days of priming, both in the number of rebound LAK cells generated and in therapeutic efficacy. The question of which regimen has greater antitumor efficacy can be answered only in a randomized study. In the meantime, attempts are ongoing to improve the efficacy of each regimen.

INTRAPERITONEAL LAK PLUS INTERLEUKIN THERAPY

In vitro studies have shown that large granular lymphocytes (LGL) incubated with IL-2 to generate LAK cells produce 10% lysis of the ovarian carcinoma cell line NIH:OVCAR-3 at an effector/target ratio as low as 3:1 and over 60% lysis at effector/target ratios of 33:1. Animal studies have shown that IL-2–incubated LGL cells or T cells can significantly prolong the median survival of a mouse NIH:OVCAR-3 model (30–32). In addition, the animal studies show that approximately 80% of LGL and T-cell populations injected directly into the peritoneal cavity remain there at 24 hr after injection, suggesting that high concentrations of locally instilled IL-2 and LAK cells can be maintained. All of these factors suggest that intraperitoneal treatment with IL-2 plus LAK therapy for ovarian carcinoma might be particularly effective.

In an attempt to deliver high concentrations of IL-2 plus activated cells locally directly to tumor sites, while minimizing systemic toxicity, the Clinical Research Branch of the Biological Response Modifiers Program of the NCI has begun studying the intraperitoneal administration of LAK cells plus IL-2 in patients with peritoneal carcinomatosis from ovarian and colon cancer (33).

The therapeutic protocol for this regimen is as follows: Three days of priming with IL-2 (100,000 units/kg) every 8 hr (total nine doses) followed by 2 days of no treatment and then 5 days of leukapheresis, with the cells from each day being incubated in IL-2 for 7 days to generate LAK cells. The cells are given back intraperitoneally daily from days 13 through 18 along with intraperitoneal IL-2 at a dose of 25,000 units/kg given every 8 hr (total of 15 doses of IL-2). On days 19 and 20, no therapy is given. Then the patient is again leukapheresed for 5 days (with the cells again incubating with IL-2 for 7 days), followed by 2 days of no treatment. The patient then receives 5 days of daily intraperitoneal LAK cells plus intraperitoneal IL-2 at the same doses given previously. The total treatment cycle, therefore, takes approximately 5 weeks.

The toxic effects of this regimen are qualitatively similar to those seen with systemic IL-2 plus LAK therapy (see Table 1). However, they are quantitatively significantly less severe. Renal dysfunction and hypotension are much milder and more easily controlled. No severe respiratory distress, myocardial ischemia or infarction, or deaths have occurred in the over 20 patients treated with this regimen. However, several patients have had Tenckhoff catheter infections, and one patient had to stop therapy because of a bowel perforation. An unexpected toxicity was the development of intraperitoneal fibrosis in a significant percentage of patients, with pseudocyst formation in the peritoneal cavity in some of these patients.

Preliminary results of treatment show partial responses in approximately 22% of patients with ovarian cancer refractory to multiple previous treatments and in 45% of patients with colon cancer (Table 4). Of note, pharmacokinetic studies do show prolonged elevation of IL-2 levels in the peritoneal cavity (with much lower systemic levels) and prolonged maintenance of LAK activity in vivo as well as interferon-γ production, indicating the favorable kinetics of locally administered IL-2 plus LAK cell therapy (34). Although still preliminary, this approach would appear to hold some promise.

Other attempts at local therapy with IL-2 plus LAK cells include three patients reported by West et al. (29) with adenocarcinoma of the lung with a malignant pleural effusion, and intraperitoneal ovarian cancer, or mesothelioma, respectively, who had clinical responses to intracavitary treatment; the report of some local antitumor responses when intralesional injections of nonrecombinant IL-2–expanded peripheral blood mononuclear cells (at a dose of 40–400 million cells per injection) were injected into subcutaneous melanoma or breast metastases (35); the use of IL-2 plus LAK cells locally in a Phase I study treating malignant gliomas (36); and direct intraarterial infusion of indium-labeled

Table 4 Intraperitoneal LAK + IL-2 Results

	Pr[a]	Cr[b]	Overall
Ovarian cancer	22	0	22
Colorectal cancer	36	9	45

[a]Pr, partial response (greater than 50% decrease in measurable tumor).
[b]Cr, complete response (no evidence of measurable tumor).
Source: Update of Steis et al. (Ref. 33).

LAK cells into renal arteries of patients with renal cell cancer, which showed that although these cells did migrate throughout the body, high local concentrations of radioactivity did remain in the kidneys at 48 hr, suggesting that reasonable prolonged local concentrations of LAK cells given intra-arterially might be feasible (37).

INTERLEUKIN-2 INFUSIONS ALONE AS AN ANTINEOPLASTIC BIOLOGICAL

As mentioned previously, even though the animal studies suggest that combined IL-2 plus LAK therapy has greater antitumor efficacy than IL-2 alone, at the highest doses of IL-2, the efficacy of IL-2 alone approaches that of the combination. Even though these are higher equivalent doses of IL-2 than can be achieved in humans because of toxicity, there still remains the question of whether exogenously generated LAK cells or the adoptive component of this therapy add anything to the antitumor therapeutic efficacy of high-dose IL-2 alone in humans. At low doses (up to 10^6 units/ m^2) of IL-2 given as a daily bolus, responses are uncommon (38). However, higher doses of IL-2 alone, in the range of those used as part of LAK therapy, given either by bolus injection or by continuous infusion, have induced responses. Using doses and a schedule of IL-2 similar to that used for systemic bolus IL-2 plus LAK therapy, one complete response was seen in 21 patients with renal cell cancer and five partial responses were seen in 16 patients with melanoma (14). A partial response was seen in a patient with renal cell cancer treated with continuous infusion of IL-2 at a dose of 3×10^6 units/m^2 daily for 4 days each week for 4 consecutive weeks (39). Although clearly preliminary, these reports indicate that IL-2 alone at high doses does have antitumor efficacy. Further studies using IL-2 alone against a variety of tumor types, but especially renal cell carcinoma and melanoma, are now being tested at a variety of treatment centers. One of the current important trials to study this question is a randomized study of high-dose IL-2 alone versus IL-2 plus LAK, in an attempt to define how critical the adoptive transfer of cells mediating LAK activity is to the antitumor response.

FURTHER QUESTIONS ABOUT LAK PLUS INTERLEUKIN-2 THERAPY

Another important consideration in both understanding the mechanism by which IL-2 induces its antitumor immune response and in learning how to optimize therapy with IL-2 is the question of which cells, or combinations of cells, are actually mediating the antitumor effect. Although cells that mediate the LAK phenomenon are the major ones expanded in IL-2–cultured peripheral blood cells from IL-2–primed patients, there are many other cells of the immune system that can respond to IL-2 and that might mediate or contribute to the antitumor response. Other IL-2–induced lymphokines, such as interferon-γ, may also be important. It has been difficult to do trafficking studies with LAK cells in patients (because of the problems of maintaining viability of radiolabeled cells), but the one study that did look at this, suggested trafficking primarily to lung, liver, and spleen. In addition, although pretreatment lymphocyte counts and post–IL-2 rebound lymphocytosis appear to correlate with response, the absolute number of LAK cells infused does not, raising the possibility that cells activated in vivo or lymphokine products of IL-2–activated cells may be a critical component of this therapy (14,20,29,39).

In an attempt to define which cells are responsible for IL-2's antitumor efficacy, biopsies of lesions before and after therapy were obtained from a patient responding to IL-2 plus LAK cell therapy (40). Before therapy, very few infiltrating cells were seen at the tumor site. After therapy, a marked infiltration with pleomorphic-appearing cells was seen. Very few of these cells expressed the NK Leu11 marker. The most common cells were $T8^+$ with a smaller number of $T4^+$ cells and monocytes. Posttreatment HLA-DR antigens were increased on both tumor as well as tumor-infiltrating cells. This suggests that $T8^+$ cytotoxic T cells, which specifically recognize tumor cells, may play an important role in mediating the antitumor response and that possibly a lymphokine (such as by interferon-γ)-induced expression of HLA-DR on tumor cells may also play a role. Clearly, many more patients need to be studied before any definite conclusions can be drawn.

In addition to the question of what role LAK (or other adoptively transferred cells) are playing in the antitumor response, there are still many questions that remain to be answered about this treatment modality. There are many potential modifications that might increase the effectiveness, including a variety of modifications in the method of growth of the cells in vitro, that have been shown to increase cell yields or increase the cytotoxic activity of the cells. Are there improvements that can be made in dose, schedule, or route of administration of the IL-2 that might both enhance therapeutic activity while also decreasing toxicity? Might IL-2 or IL-2 plus LAK be particularly effective in the adjuvant setting when the host tumor burden is low, as would be predicted from the many animal studies that suggest that immunotherapy is most effective when host tumor burden is small? What will the overall response rate be in non-Hodgkin's lymphomas, and are there other tumor types not yet treated with IL-2 plus LAK that might be responsive? Would maintenance therapy or repeated treatment cycles yield more durable responses? Are there other factors, such as correction of the severe zinc, ascorbic acid, or vitamin B_6 depletions that occur in patients treated with IL-2 plus LAK, that might improve the cytotoxicity of the regimen (41)? There still remains a question of whether a serum IL-2 inhibitor, or inhibitors, such as has been found in mice and in human serum, may play a role in the lack of response in certain patients and, if so, whether or not approaches to inhibit or remove this inhibitor might enhance IL-2's therapeutic effects (42,43). Combinations with other treatment modalities including chemotherapeutic agents, such as cyclophosphamide (Cytoxan), both to inhibit suppressor cell function as well as possible synergistic cytotoxicity; or doxorubicin (Adriamycin; which has been shown to have synergistic activity with IL-2 plus LAK against an animal tumor model), or other biological response modifiers, such as interferons, which have shown synergistic activity in vitro in combination with IL-2, are both ongoing and planned (44–47).

Another approach, which uses adoptively transferred cells, is the isolation of tumor infiltrating lymphocytes and their expansion with IL-2. Although a variety of immune cells infiltrate tumor sites, the most common lymphocytes that can be isolated and expanded from tumor sites are $T8^+$ cytotoxic lymphocytes, which recognize tumor-associated antigen as well as self-MHC gene products and are thus specific for the particular tumor from which they are isolated. These have been reported to have 50 to 100 times greater activity than do LAK cells against the specific tumor from which they were derived (42). It is hoped that these can be given with lower doses of IL-2, thus reducing toxicity, while still retaining high antitumor activity. However, major technical difficulties in isolating these cells from tumors and expanding them in vivo still remain, and only future clinical trials will answer how effective this approach might be.

CONCLUSION

In summary, preliminary studies have shown activity for IL-2 alone or combined IL-2 plus LAK cell therapy in the treatment of various malignancies but, especially, melanoma, renal cell carcinoma, low-grade non-Hodgkin's lymphomas and, to a lesser extent, against colon cancer. This is encouraging because melanoma and renal cell carinoma are malignancies that have low response rates to all therapies yet attempted, and the adoptive transfer of LAK cells and IL-2 is a new form of immunotherapy about which much still needs to be learned to maximize its efficacy. Very preliminary results have shown an overall response rate of approximately 15% to 33% against renal cell carcinoma (overall present combined response rate of 26%) and 21% against melanoma. Although these responses are encouraging, clearly further improvements in therapeutic efficacy are needed if IL-2 plus LAK therapy is to prove beneficial for most patients with metastatic renal cell carcinoma or melanoma. In addition, the high-dose intravenous bolus regimen, as it is currently given, has moderately severe toxicity. Attempts are ongoing in the laboratory and through a variety of clinical trials to improve the effectiveness of this therapy, to decrease toxicity, and to combine it with other treatment modalities to help establish what role IL-2 alone or in combination with LAK cells might have in the overall treatment of cancer.

REFERENCES

1. Rosenberg S A, Lotze M T, Muul L M, et al. Special Report. Observations on the systemic administration of autologous lymphokine-activated killer cells and recombinant interleukin-2 to patients with metastatic cancer. N Engl J Med 1985;313:1485–1491.
2. Burdach S, Shatsky M, Wagenhorts B, et al. Receptor-specific modulation of myelopoiesis by recombinant DNA-derived interleukin-2. J Immunol 1987;139:452–458.
3. Wahl S M, McCartney-Francis N, Hunt D A, et al. Monocyte interleukin-2 receptor gene expression and interleukin-2 augmentation of microbicidal activity. J Immunol 1987; 139:1342–1347.
4. Grimm E A, Mazumder A, Zhang H Z, et al. Lymphokine-activated killer cell phenomenon: Lysis of natural killer-resistant fresh solid tumor cells by interleukin-2–activated autologous human peripheral blood lymphocytes. J Exp Med 1982;155:1823–1841.
5. Ortaldo J R, Mason A, Overton R. Lymphokine-activated killer cells. Analysis of progenitors and effectors. J Exp Med 1986;164:1193–1205.
6. Kyogo I, Tilden A B, Kumagai K, et al. Leu11$^+$ lymphocytes with natural killer (NK) activity are precursors of recombinant interleukin-2 (rIL-2)-induced activated killer (AK) cells. J Immunol 1985;134:802.
7. Lanier L L, Phillips J H. Human thymic and peripheral blood non-MHC-restricted cytotoxic lymphocytes. Med Oncol Tumor Pharmacother 1986;3:247–254.
8. Phillips J H, Lanier L L. Dissection of the lymphokine-acti-

vated killer phenomenon relative contribution of peripheral blood natural killer cells and T lymphocytes to cytolysis. J Exp Med 1986; 164:814–825.
9. Kallard T, Belfrage H, Bhiladvala P, et al. Analysis of the murine lymphokine-activated killer (LAK) cell phenomenon: Dissection of effectors and progenitors into NK- and T-like cells. J Immunol 1987; 138:3640–3645.
10. Danile N K, Doyle L V, Bradley E C. Interleukin-2 activated human killer cells are derived from phenotypically heterogeneous precursors. J Immunol 1986; 137:2814–2822.
11. Mule J J, Shu S, Schwarz S L, et al. Adoptive immunotherapy of established pulmonary metastases with LAK cells and recombinant interleukin-2. Science 1984; 225:1487–1489.
12. Lafreniere R, Rosenberg S A. Successful immunotherapy of murine experimental hepatic metastases with lymphokine-activated killer cells and recombinant interleukin-2. Cancer Res 1985; 45:3735–3741.
13. Bubenik J, Indrova M. The anti-tumour efficacy of human recombinant interleukin-2. Cancer Immunol Immunother 1987; 24:269–271.
14. Rosenberg S A, Lotze M T, Muul L M, et al. A progress report on the treatment of 157 patients with advanced cancer using lymphokine-activated killer cells and interleukin-2 or high-dose interleukin-2 alone. N Engl J Med 1987; 316:889–897.
15. Braun D P, Harris J E. Abnormal indomethacin-sensitive suppression in peripheral blood mononuclear cells of cancer patients restricts augmentation by interleukin-2. J Biol Response Mod 1986; 3:533–540.
16. Griscold D E, Alessi S, Badger A M, et al. Inhibition of T suppressor cell expression by histamine type 2 (H_2) receptor antagonists. J Immunol 1984; 132:3054–3057.
17. Holda J H, Maier T, Claman, H N. Natural suppressor activity in graft-vs-host spleen and normal bone marrow is augmented by IL-2 and interferon-gamma. J Immunol 1986; 137:3538–3543.

18. Jones G R N. Cancer therapy: Phenothiazine in an unexpected role. Tumour 1985; 71:563–569.
19. Margolin K, Jaffe H S, Atkins M B, et al. Toxicity of interleukin-2 and lymphokine-activated killer cell therapy. Proc ASCO 1987; 6:251.
20. Educational booklet. 23rd Annual Meeting ASCO. p. 19–21. May 17–19, 1987.
21. Belldegrun A, Webb D E, Austin H A, et al. Effects of interleukin-2 on renal function in patients receiving immunotherapy for advanced cancer. Ann Intern Med 1987; 106: 817–822.
22. Ettinghausen S E, Moore J G, White D E, et al. Hematologic effects of immunotherapy with lymphokine-activated killer cells and recombinant interleukin-2 in cancer patients. Blood 1987; 69:1654–1660.
23. Orr D, Yannelli J, Sharp E, et al. Eosinophilia during interleukin-2–activated killer cell therapy. Proc ASCO 1987; 6: 247.
24. Vetto J T, Papa M Z, Lotze M T, et al. Reduction of toxicity of interleukin-2 and lymphokine-activated killer cells in humans by the administration of corticosteroids. J Clin Oncol 1987; 5:496–503.
25. Parkinson D R, Snydman D R, Weisfuse B, et al. Hepatitis A (HAV) infection occurring following IL-2/LAK cell therapy. Proc ASCO 1987; 6:234.
26. Beckner S K, Maluish A E, Longo D L, et al. Lymphokine-activated killer cells: Culture conditions for the generation of maximal in vitro cytotoxicity in cells for normal donors. Cancer Res 1987; 47:5504–5508.
27. Fisher R L, Coltman C A, Doroshow J H, et al. Phase II clinical trial of interleukin II plus lymphokine activated killer cells in metastatic renal cancer. Proc ASCO 1987; 6:244.
28. Dutcher J P, Creekmore S, Weiss, G R, et al. Phase II study of high dose interleukin-2 and lymphokine activated killer cells in patients with melanoma. Proc ASCO 1987; 6:246.
29. West W H, Tauer K W, Yannelli J R, et al. Constant-infusion recombinant interleukin-2 in adoptive immunotherapy of advanced cancer. N Engl J Med 1987; 316:898–905.

30. Ortaldo J R, Porter H R, Miller P, et al. Adoptive cellular immunotherapy of human ovarian carcinoma xenografts in nude mice. Cancer Res 1986; 46:4414-4419.
31. Hamilton T C, Ozols R F, Longo D L. Biologic therapy for the treatment of malignant common epithelial tumors of the ovary. Cancer 1987; 60:2054-2063.
32. Hamilton T C, Ozols R F, Young R C, et al. Malignant common epithelial tumors of the ovary. (In preparation).
33. Steis R, Bookman M, Clark J, et al. Intraperitoneal lymphokine activated killer cell and interleukin-2 therapy for peritoneal carcinomatosis: Toxicity, efficacy, and laboratory results. Proc ASCO 1987; 6:250.
34. Rossio J L, Rager H, Rice R, et al. IL-2 levels and in vivo induction of gamma interferon in ascites and sera of cancer patients during intraperitoneal LAK cell-IL-2 therapy. Proc ASCO 1987; 6:250.
35. Adler A, Stein J A, Kedar E, et al. Intralesional injection of interleukin-2-expanded autologous lymphocytes in melanoma and breast cancer patients: A pilot study. J Biol Response Mod 1984; 3:491-500.
36. Jacobs S K, Wilson D. J, Kornblith P L, et al. Interleukin-2 or autologous lymphokine-activated killer cell treatment of malignant glioma: Phase I trial. Cancer Res 1986; 46:2101-2104.
37. Morita T, Yonese Y, Minato N. In vivo distribution of recombinant interleukin-2-activated autologous lymphocytes administered by intra-arterial infusion in patients with renal cell carcinoma. J Natl Cancer Inst 1987; 78:441-447.
38. Atkins M B, Gould J A, Allegretta M, et al. Phase I evaluation of recombinant interleukin-2 in patients with advanced malignant disease. J Clin Oncol 1986; 4:1380-1391.
39. Sondel P M, Hank J A, Kohler P C. Status and potential of interleukin-2 or the treatment of neoplastic disease. Oncology 1987; :41.
40. Longo D L. Biological therapy of cancer. (In preparation).
41. Marcus S L, Dutcher J, Ciobanu N. Micronutrient depletion in patients treated with high-dose interleukin-2 and lymphokine-activated killer cells. Proc ASCO 1987; 6:247.

42. Male D, Lelchuk R, Curry S. Serum IL-2 inhibitor in mice. II. Molecular characteristics. Immunology 1985;56:119–127.
43. Itoh K, Tilden A B, Balch C M. Role of interleukin-2 and a serum suppressive factor on the induction of activated killer cells cytotoxic for autologous human melanoma cells. Cancer Res 1985;45:3173–3178.
44. Salup R R, Back T C, Wiltrout R H. Successful treatment of advanced murine renal cell cancer by bicompartmental adoptive chemoimmunotherapy. J Immunol 1987;138:641–647.
45. Krigel R, Poiesz B, Comis R, et al. A Phase I study of recombinant interleukin-2 plus recombinant beta-ser 17 interferon. Proc ASCO 1986;5:225.
46. Kolitz J E, Merluzzi V J, Welte K, et al. A phase I trial of recombinant interleukin-2 and cyclophosphamide in advanced malignancy. Proc ASCO 1986;5:235.
47. Rosenberg S A, Spiess P, Lafreniere R. A new approach to the adoptive immunotherapy of cancer with tumor-infiltrating lymphocytes. Science 1986;233:1318–1321.

4

Lymphokine-Activated Killer Lymphocytes: Extramural Clinical Trials

IRENA J. SNIECINSKI
City of Hope National Medical Center, Duarte, California

During the last 8 years, a substantial effort has been focused on development of the new immunotherapeutic regimens for the treatment of some highly metastatic, otherwise nonresponsive, human tumors (1). Adoptive transfer of in vitro lymphokine-activated killer cells (LAK) in combination with in vivo administration of recombinant interleukin-2 (rIL-2) has resulted in regression of advanced neoplastic diseases in several murine tumor systems (2,3). Adoptive cellular immunotherapy (ACI) of human cancers using autologous LAK cells and rIL-2 has been undertaken by the Surgery Branch of the National Cancer Institute (NCI) based on these promising results. In December 1985, Rosenberg and colleagues reported the results of this therapy in 25 patients with advanced cancer (4). In February 1986, the NCI trial was extended to six extramural cancer centers including the University of California, San Francisco Medical Center; City of Hope National Medical Center; Loyola University Medical Center; University of Texas,

Table 1 US Extramural IL-2-LAK Working Group Treatment Protocols

Bolus IL-2 and LAK cells in renal carcinoma, melanoma, and colorectal carcinoma
Hybrid IL-2 and LAK cells in renal carcinoma and melanoma
Continuous infusion IL-2 and LAK cells in renal carcinoma and melanoma
Randomized study; bolus vs infusion IL-2 and LAK cells in renal carcinoma
Bolus IL-2 and LAK cells in other malignancies
Bolus IL-2 alone in malignant melanoma

San Antonio Health Sciences Center; Albert Einstein–Montefiore Medical Center, and Tufts–New England Medical Center.

The initial objectives of these NCI extramural Phase II trials were to determine whether the complex laboratory and clinical technology involved in the treatment with IL-2 and LAK cells could be administered in other cancer centers and to confirm its efficacy. Recently, 24 additional cancer centers were included in the investigation of this type of therapy. Numerous treatment protocols have been designed in an effort to maximize the efficacy and minimize the toxicity of the ACI therapy (Table 1). Also, randomized trials have been developed to compare the benefit of high-dose IL-2 alone or with that of IL-2 with LAK cells.

PROTOCOLS

The initial regimen developed at the NCI in piloting studies consisted of 16 days of treatment divided into three phases (Fig. 1). During the priming phase, patients received for 5 days the bolus infusion of IL-2 at a dose of 100,000 units/kg every 8 hr. Forty-eight hours after the last dose of IL-2, patients were subjected to five daily leukaphereses of 4-hr duration. The lymphocytes were isolated and cultured for 3 to 4 days to amplify the LAK activity. After last leukapheresis, patient cells were reinfused along with additional boluses of IL-2 during the last 4 days to promote con-

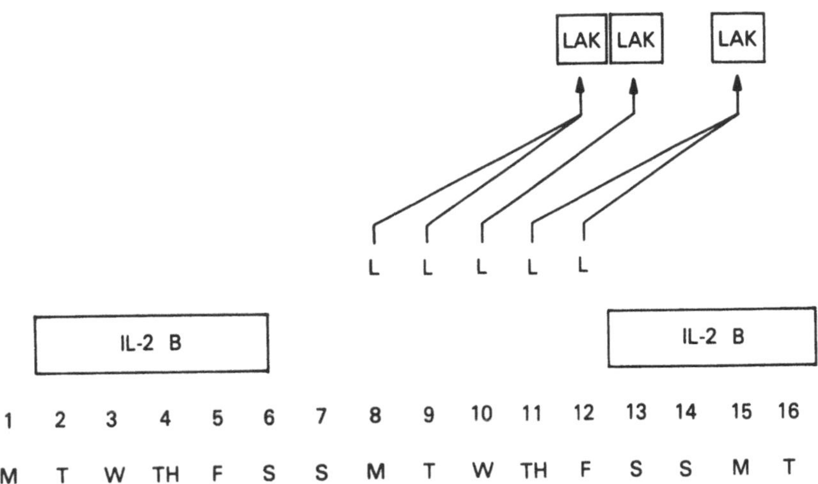

Figure 1 Bolus IL-2 LAK protocol.

tinued LAK activity in vivo. This protocol was subsequently modified in an attempt to improve antitumor responses while decreasing the toxicity of this treatment. In modified "hybrid" protocol, the priming with IL-2 was decreased from 5 to 3 days to decrease the toxicity (Fig. 2). The leukapheresis was started 24 hr after the last IL-2-priming dose to maximize cell yields. The number of leukapheresis procedures was decreased to four, each continued for 5 hr. The IL-2 infused along with LAK cells was given by continuous infusion to prolong the duration of LAK activity in vivo. Subsequently, the "continuous infusion" protocol was introduced to improve the number of LAK cells and the antitumor efficacy of the hybrid therapy. In this protocol, the IL-2 used in priming and with LAK cells was given by continuous infusion at similar doses (Fig. 3). Priming was done for 5 days, followed by a 4-day leukapheresis, and then by reinfusion of LAK cells along with IL-2 administration for 5 days.

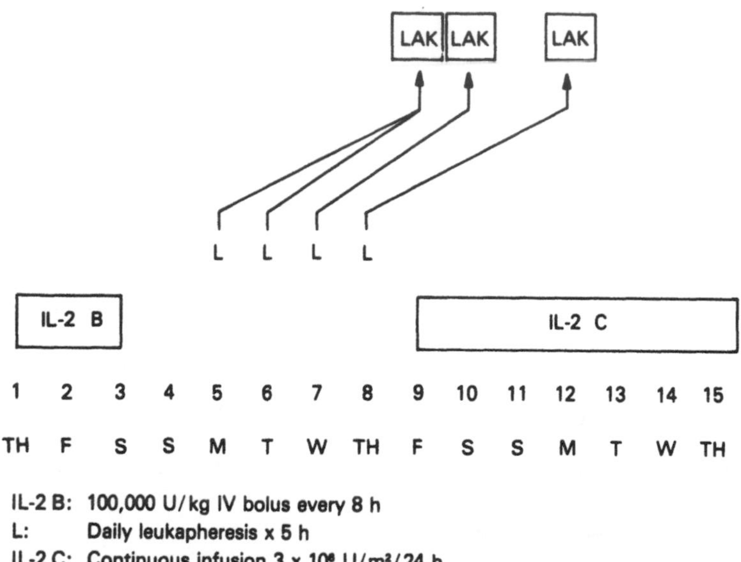

Figure 2 Hybrid IL-2/LAK protocol.

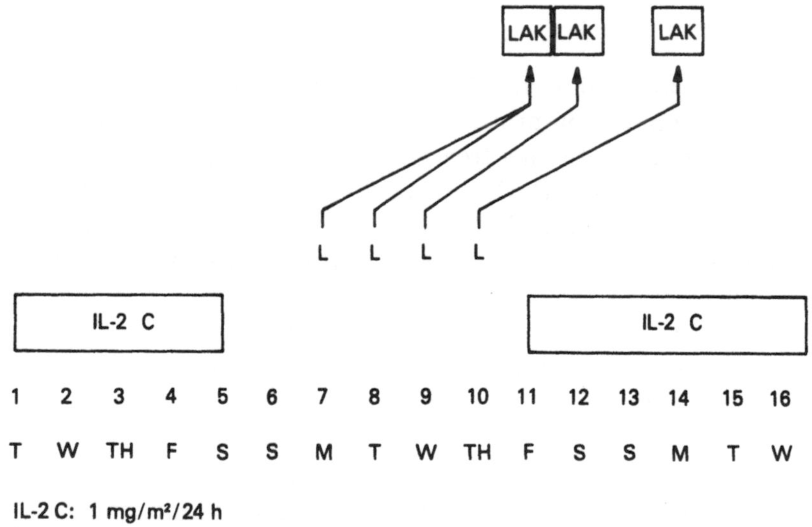

Figure 3 Modified IL-2/LAK protocol (continuous IL-2 infusion).

CELL COLLECTION AND PREPARATION

A number of the strategies have been developed for the collection of purified mononuclear subset of leukocytes and the in vitro activation of the harvested cells. Attention was given to the leukapheresis parameters to increase the efficiency and purity of mononuclear cells subset collection while eliminating other, possibly antagonistic, subsets. Initially, the leukapheresis procedures were performed using "low" centrifugal acceleration of 1010 rpm, the blood flow rate of 45 ml/min, and the collect rate of 2.1 ml/min. More recently, the settings on the leukapheresis protocol were modified to include an increased rpm of 1400, the blood flow rates of 75 ml/min, and the collect rate of 3.7 ml/min (Table 2). The cells were collected for the first 4 hr at an interface hematocrit of 2%, and during the last hour at the hematocrit of 4%. These modifications increased the cell yields and purity of mononuclear cells (Table 3).

Mononuclear cell yields on each of the 4 days of apheresis correlated positively with rebound lymphocytosis. The highest mononuclear cell yields and rebound peripheral lymphocyte counts were seen on day 1 of leukapheresis in patients primed with a 5-day infusion of IL-2. Patients enrolled in a 3-day IL-2–priming protocol had the highest mononuclear cell yields and peripheral rebound lymphocytosis on day 2 of leukapheresis. Overall, those

Table 2 IL-2-LAK Leukapheresis Parameters

Procedure characteristics	High rpm	Low rpm
Blood flow rate (ml/min)	75 (70–80)	45 (40–50)
Collection rate (ml/min)	3.7 (3.5–4.0)	2.1 (1.8–2.2)
RPM	1400	1010
WBC/ACD ratio (ml)	13–18	11–15
Time (hr)	5	5
Volume processed (L)	21 (14–24)	12 (11–13)

Table 3 IL-2-LAK Trial Characteristics of Mononuclear Cell Concentrates

Characteristics	High rpm	Low rpm
Volume (ml)	1100 (860–1400)	570 (450–650)
Total mononuclear leukocytes × 10^{10}	6.3 (1.8–15.0)	3.3 (0.9–10.0)
% mononuclear leukocytes	94 (88–100)	87 (50–100)
Hct%	3.0 (2.4–4.3)	4.4 (3.2–6.0)
platelets × 10^3 ml	529 (392–800)	825 (680–998)
Total platelets × 10^{11}	5.8 (4.3–8.8)	4.7 (3.9–5.7)
Total mononuclear leukocytes per patient	25.3 (11.5–34.6)	13.3 (6.1–26.1)

patients who were primed with IL-2 infusion had the highest rebound of peripheral lymphocyte count and the highest mononuclear cell yields. The highest mononuclear cell yields and rebound lymphocyte counts were seen on day 1 of leukapheresis, and declined steadily thereafter. Patients treated with a 5-day IL-2 had higher cell yields and peripheral lymphocytosis than those treated with a 3-day bolus of IL-2.

Unfractionated mononuclear cells must undergo several processing steps before reinfusion into patients. The initial protocol was labor-intensive and carried risk of microbial contamination (5). Routinely, cells were purified on a Hypaque–Ficoll gradient, pooled, washed two times, and cultured in 2.3 L roller bottles. After 3 to 4 days of culture, cells were manipulated through numerous steps of pooling, washing, concentrating, and resuspending before infusion into the patient (Table 4).

Subsequently, the complexity of the manual process was reduced to make it easily acceptable for routine use. The Ficoll–Hypaque step was eliminated, thus increasing the total cell yield as well as the specific activity. Automated LAK cell-processing and harvesting have been developed using the blood cell separators (Table 5) (6). Standard polyolefin platelet storage bags have been adapted for long-term culture of LAK cells in IL-2, thus allowing the whole process to be performed in a closed-system manner. In addition, the serum-free medium replaced the human serum. These modifications decreased the degree of technical manipulation, reduced the cost, and eliminated the potential sources of microbial contamination. The results of our study in which the automated technique was used to process the mononuclear cell concentrate and harvest LAK cells out of cultures are presented in Table 6. Omission of Ficoll–Hypaque gradient increased the mononuclear cell recovery after processing and harvesting of cells. Mean mononuclear cell recovery after automated processing without the Ficoll–Hypaque gradient and automated harvest (68%) was comparable with that obtained by manual processing and harvest without the Ficoll–Hypaque gradient (64%). These recoveries were significantly higher than the mononuclear cell recoveries using manual technique of processing with Ficoll–Hypaque gradient and

Table 4 Manual Processing and Harvest of LAK Cells

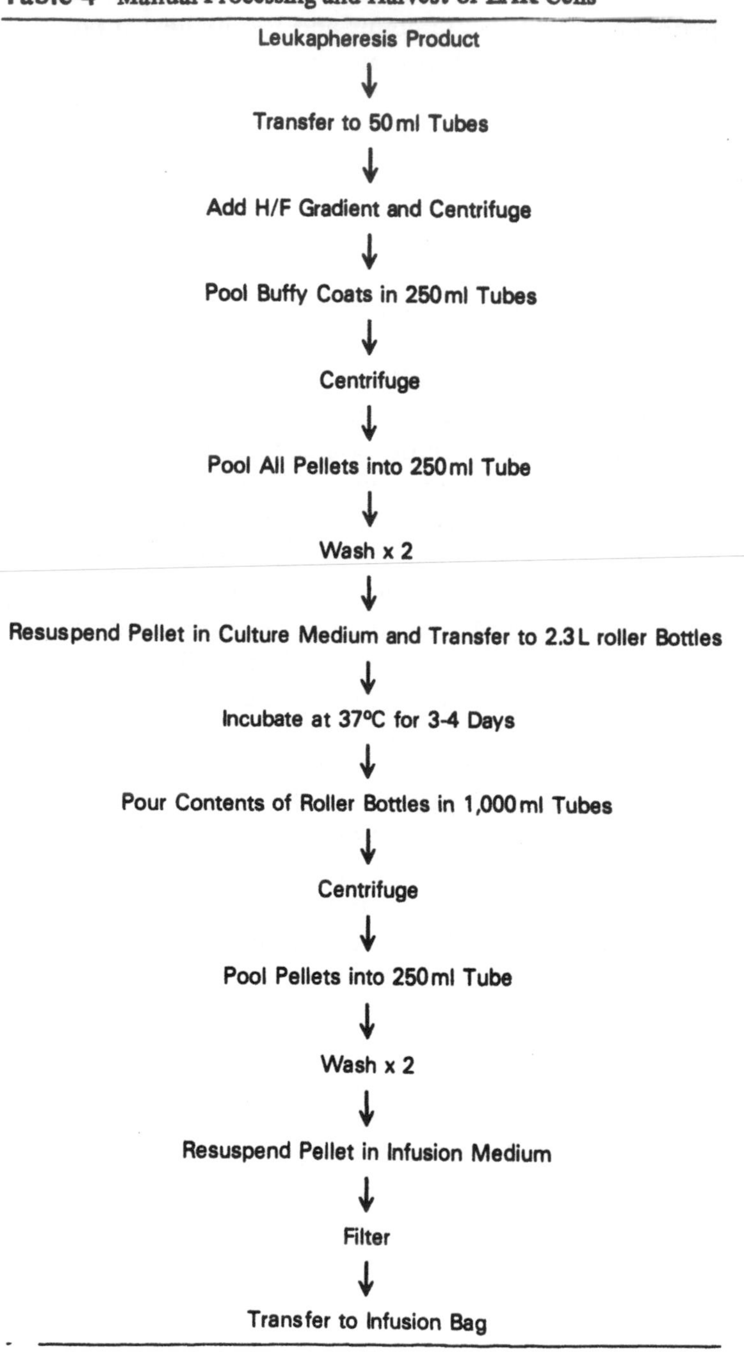

Table 5 Automated Processing and Harvest of LAK Cells

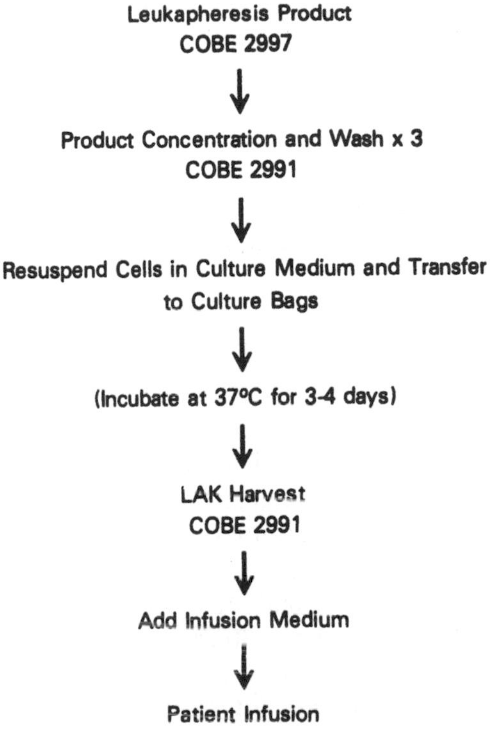

Leukapheresis Product
COBE 2997

↓

Product Concentration and Wash x 3
COBE 2991

↓

Resuspend Cells in Culture Medium and Transfer to Culture Bags

↓

(Incubate at 37°C for 3-4 days)

↓

LAK Harvest
COBE 2991

↓

Add Infusion Medium

↓

Patient Infusion

Table 6 Cell Recoveries: Methods of Processing and Harvesting

	Manual and Hypaque-Ficoll (H-F)	Manual: no H-F	Automated: no H-F
Infused cell recovery (%) from leukapheresis (after processing and harvesting)	41	64	68
Average loss (%)			
from processing before culture	24	9	8
from harvesting after culture	35	27	24

manual harvest technique (41%). Average loss of mononuclear cells during processing was 24% for the manual technique with the Ficoll–Hypaque gradient, 9% for the manual technique without the Ficoll–Hypaque gradient, and 8% for the automated technique without the gradient. Manual Ficoll–Hypaque processing involved more than 160 breaks in sterility, and necessitated use of a laminar flow hood: automated Ficoll–Hypaque processing required less than ten interruptions in the closed-system apheresis kit. Cell viability after both manual and automated processing was more than 90% in all cases.

PATIENT SELECTION

All patients accepted for treatment with IL-2–LAK cells had clearly progressive and measurable metastatic malignancy that was refractory to standard therapy. They were ambulatory with an Eastern Cooperative Oncology Group performance status 0 or 1 and appropriate laboratory evidence of normal renal, liver, bone marrow, and heart function. Other eligibility criteria are listed in Table 7. Of importance is the criterion of prior immunotherapy with agents other than tumor necrosis factor, interferon-γ or interleukin. Prior immunotherapy should have been completed at least 4 weeks before beginning IL-2–LAK treatment. Concurrent chemotherapy, radiotherapy, or immunotherapy was not permitted.

Treatment with IL-2 and administration of LAK were performed in an intensive care unit with central venous pressure monitoring (7). During IL-2 administration, patients also received acetaminophen, 650 mg, orally every 4 hr and indomethacin, 25 mg, orally every 6 hr for control of fever; ranitidine or cimetidine for prophylaxis of gastrointestinal bleeding; hydroxyzine hydrochloride or diphenhydramine for treatment of rash and pruritus; and meperidine, 25 to 50 mg, intravenously to control chills and rigors. Steroids were administered only in life-threatening situations.

Patients were evaluated for response at the completion of therapy and at intervals of 4 weeks thereafter (8). A response was considered to be complete if all measurable tumor disappeared. Partial response was defined as a 50% or greater reduction of the sum of the product of the longest perpendicular diameters of all measurable lesions with no new lesions developing. Stable disease included any reduction of less than 50% or any increase of less than 25%. Progressive disease included any increase of 25% or greater. Patients with evidence of response were retreated with autologous LAK cells for an additional two cyles at 3-month intervals.

Ninety-three patients completed the bolus protocol, 73 patients completed the hybrid protocol and 95 patients were accrued to the infusion protocol. Scheduled doses of IL-2 were administered either at full dose or omitted, depending on patient toxicity. Pa-

Table 7 Eligibility Criteria for IL-2-LAK Study

Advanced measurable disease that is clearly progressive
Good performance status (ECOG PS = 0 or 1; Karnofsky PS $\geqslant$ 80%)
Age $\geqslant$ 16
WBC $\geqslant 3500/mm^3$, platelet count $\geqslant 100{,}000/mm^3$
Bilirubin $\leqslant$ 1.5 mg/dl
Creatinine $\leqslant$ 1.5 mg/dl
FEV_1 $\leqslant$ 75% of predicted
Normal stress test and ECG; no evidence of previous MI, CHF, CAD, or serious cardiac arrhythmia
Normal brain scan
Absence of HBsAg and HIV antibodies
Concurrent chemo/radio/immunotherapy not permitted
Prior immunotherapy limited to monoclonal antibodies, interferon-γ or -β, and tumor vaccines

tients enrolled in the bolus protocol received 93% of the priming IL-2 and 71% of IL-2 administered during LAK cell infusion. The mean number of LAK cells reinfused was 7.6×10^{10}. Patients receiving treatment according to hybrid protocol received 93% of the priming IL-2 and 83% of therapeutic IL-2, with a mean number of 8.5×10^{10} LAK cells. The infusion group received the highest dose of the priming and therapeutic IL-2, 98% and 85%, respectively, and the highest mean number of LAK cells (13.6×10^{10}).

RESULTS

The clinical responses in patients with renal carcinoma are summarized in Table 8 in which the therapeutic efficacy of various treatment protocols is compared (9,10). It appears that a 5-day priming with IL-2 whether by bolus or continuous infusion is more effective than a 3-day priming. The response rates for a 5-day bolus and 5-day infusion were 16% and 35%, respectively, compared with 6% or a hybrid protocol. Similar correlations were observed in patients with malignant melanoma (Table 9). In this

Table 8 IL-2-LAK Extramural Trials: Clinical Responses to IL-2-LAK Therapy in Patients with Metastatic Renal Carcinoma

Treatment modality	Evaluable patients	CR rate	PR rate	Response rate (%)
Bolus IL-2 and LAK cells	32	2	3	16
Hybrid IL-2 and LAK cells	32	0	2	6
Continuous infusion IL-2 and LAK cells	23	2	6	35

Table 9 IL-2-LAK Extramural Trials: Clinical Responses to IL-2-LAK Therapy in Patients with Metastatic Melanoma

Treatment plan	Evaluable patients	CR rate	PR rate	Response rate (%)
Bolus IL-2 and LAK cells	32	1	5	19
Hybrid IL-2 and LAK cells	42	0	6	14
Continuous infusion IL-2 and LAK cells	9	NA	NA	20

NA, too early to evaluate.

group, the response rate for the bolus and infusion protocols were 19% and 20% versus 14% for a hybrid protocol. At the present time, it is too early to know which of the two 5-day regimens, bolus or infusion protocol, has greater antitumor efficacy. A randomized study comparing the efficacy of these two regimens in patients with renal cell carcinoma is being performed at the extramural centers.

To define the antitumor efficacy of IL-2 alone, the extramural centers initiated the randomized study in patients with malignant melanomas using doses and schedules of IL-2 similar to that used for systemic bolus IL-2 and LAK regimen (8). Objective partial responses were seen in 19% (5/26) patients who could be evaluated, with three (14%) pathologically confirmed complete responses. The overall response rate in patients with malignant melanoma was 31%, which is comparable with that reported earlier by the NCI (Table 10). Although clearly preliminary, these reports indicate that IL-2 alone at high doses does have antitumor efficacy. A randomized study of IL-2 versus IL-2 combined with LAK cells is now being done to define how critical the LAK activity is in the antitumor response.

Table 10 High-Dose Interleukin IL-2 Alone in Patients with Metastatic Melanoma

Investigators	Evaluable patients	CR rate	PR rate	Overall response rate (%)
NCI trial	16	0	5	31
Extramural trial	26	3	5	31

DISCUSSION

There is an ongoing effort to identify the mononuclear cell subset that is most efficacious from the viewpoint of natural killer (NK) and cytotoxic (LAK) activity (11–13). Current methods for culturing LAK cells produce a mixed population of mononuclear cells, some of which have LAK activity, but most of which do not. This heterogeneity may both affect the activity and contribute to toxicity of the therapy. Identification and subsequent selection of appropriate mononuclear cells with a high level of killing activity may yield a much more potent population of LAK cells for clinical use. This may correlate with enhanced clinical efficacy. In addition, the ability to capture the cells that constitute the pure population of precursors of LAK cells from patient's peripheral blood will greatly simplify and reduce the cost of LAK cell preparation.

Several extramural cancer centers are involved in a study with the specific goal of characterizing the antigenic phenotype of cells isolated from the peripheral blood by leukapheresis to compare them with the phenotype of the LAK cells. Initial results from our institution failed to identify the unique subset of cells that would unequivocally be responsible for LAK activity. Both, leukapheresis concentrate and LAK cell concentrate showed similar expression of T–cell- and NK–cell-associated markers.

The IL-2–LAK therapy has been associated with a number of serious toxic reactions and the experience of the Extramural Group has confirmed earlier observations made by the NCI (7,8). In the bolus study, 56 of 83 patients who could be evaluated experienced 109 episodes of grade 3 or 4 activity. The exact mechanisms of all of the toxic reactions is not known. However, one major mechanism appears to be related to increased capillary permeability, evidenced by massive fluid shifts into extravascular tissues (14–16). The spectrum of toxic reactions observed in the extramural studies and the frequency of each are presented in Table 11. Nearly 75% of patients developed profound hypotension, requiring vasopressor agents and intensive care nursing. Over 50%

Table 11 Toxicity of Treatment with IL-2 and LAK Cells (%)

Hypotension	74
Anemia	72
Weight gain (>10%)	48
Anuria	38
Pulmonary complication	27
Cardiac arrhythmias	21
Infection	13
Thrombocytopenia	34
Cardiovascular	8
Neurological dysfunction	36
Perforated viscus	2
GI hemorrhage	2
Elevated liver function tests	61

of patients developed prerenal azotemia with anuria and increased serum creatinine levels, and over 30% of patients experienced fluid retention exceeding 10% of their body weight. Other adverse reactions included fever and chills, nausea and vomiting, fatigue, elevated liver function test results (both transaminases and bilirubin), diffuse erythematous skin rash, anemia requiring red cell transfusions, eosinophilia, and alterations in mental function. Cardiovascular toxicity was seen in 21% of patients and either angina or myocardial infarction in 8% of patients. All of these toxic reactions were of short duration and promptly reversible upon discontinuation of IL-2.

Hepatitis A infection has been seen in several of IL-2–LAK-treated patients, but this appeared to be related to a contamination of a single lot of medium used for growing of LAK cells (17). Occurrence of this complication provided a major impetus for development of the serum-free media to maintain the LAK cells in culture.

Eleven percent of patients developed catheter-related infections, necessitating the use of intravenous antibiotics. Implementation of the "sandwich" technique of catheter dressing decreased the frequency of this problem (19).

Although administration of acetaminophen, antihistamines, and other agents are somewhat effective in distinguishing the toxic effects of IL-2 therapy, the extreme morbidity associated with treatment has led to investigations of alternative methods of administering IL-2 (19). Our preliminary results indicate that the administration of IL-2 by continuous infusion was associated with the adverse effects that were qualitatively similar to, but quantitatively significantly less than, those seen with the bolus IL-2–LAK regimen.

CONCLUSIONS

The preliminary results from the extramural studies have shown a role IL-2 and LAK cells in the treatment of various malignancies but, especially, in renal cell carcinoma and malignant melanoma.

Further clinical research studies in other tumor types are underway at the cancer centers in the United States. Adoptive immunotherapy with LAK cells and IL-2 is a new form of therapy about which much still needs to be learned to maximize its efficacy. Although the overall responses to this treatment are encouraging, clearly additional improvements in the therapeutic efficacy are needed. As alluded to, several improvements have been made to reduce the complexity and expense of LAK generation. Also, the original treatment protocol underwent several modifications to maximize the yields of cytotoxic cells and to decrease the toxicity. There are many potential modifications that might enhance the tumoricidal effect of LAK therapy, including (a) improved techniques for cell collection to harvest the maximal number of purified leukocyte subsets from the peripheral blood; (b) further refinement of extracorporeal handling and culture techniques to generate sterile and highly cytotoxic cells; (c) improvement in schedule and dose of IL-2 administration; (d) use of other biological modifiers to increase the number of cytotoxic cells while decreasing the toxicity of treatment; (e) combined use of LAK cells and chemotherapeutic agents either to inhibit suppressive cell function or to enhance the cytotoxic activity; (f) isolation of tumor-infiltrating lymphocytes with subsequent expansion and transformation of these cells in long-term culture with various modifiers; and, (g) identification of more predictive clinical and laboratory assessments for evaluation of LAK cell therapeutic efficacy and toxicity.

REFERENCES

1. Rosenberg SA. Adoptive immunotherapy of cancer: Accomplishments and prospects. Cancer Treat Rep 1984; 68:233-255.
2. Mazumder A, Rosenberg SA. Successful immunotherapy by natural killer resistant established pulmonary melanoma metastases by the intravenous adoptive transfer of syngeneic lymphocytes activated in vitro by interleukin-2. J Exp Med 1984; 159:495-507.

3. Lafreniere R, Rosenberg SA. Adoptive immunotherapy of murine hepatic metastases with lymphokine-activated killer (LAK) cells and recombinant interleukin-2 (rIL-2) can mediate the regression of both immunogenic and nonimmunogenic sarcomas and an adenocarcinoma. J Immunol 1985; 135:4273–4280.
4. Rosenberg SA, Lotze MT, Muul LM, Leitman S, Chang AE, et al. Observations on the systemic administration of autologous lymphokine-activated killer cells and recombinant interleukin-2 to patients with metastatic cancer. N Engl J Med 1985;313:1485–1492.
5. Muul LM, Director EP, Hyatt CL, Rosenberg SA. Large scale production of human lymphokine activated killer cells for use in adoptive immunotherapy. J Immunol Methods 1986; 88:265–275.
6. Muul LM, Nason-Burchenal K, Carter CS, Cullis H, Slavin D, et al. Development of an automated closed system for generation of human lymphokine activated killer cells for use in adoptive immunotherapy. J Immunol Methods 1987; 101: 171–181.
7. Margolin K, Jaffe HS, Hawkins M, Atkins MB, Ciobanu N, et al. Toxicity of interleukin-2 and lymphokine activated killer cell therapy. Proc ASCO 1987;6:251.
8. Rosenberg SA, Lotze MT, Muul LM, Change AE, Avis FP, et al. A progress report on the treatment of 157 patients with advanced cancer using lymphokine-activated killer cells and interleukin-2 or high dose interleukin-2 alone. N Engl J Med 1987;316:889–897.
9. Fisher RL, Coltman CA, Doroshow JH, Rayner A, Hawkins MJ, et al. Phase II clinical trial of interleukin-II plus lymphokine activated killer cells in metastatic renal cancer. Ann Intern Med 1988;108:518–523.
10. Dutcher JR, Creekman S, Weiss GR, Margolin K, Markowitz AB, et al. Phase II study of high dose interleukin-2 and lymphokine activated killer cells in patients with melanoma. Proc ASCO 1987;6:246.

11. Itoh K, Tilden AB, Kumagai K, Balch CM. LeuII+ lymphocyte with natural killer activity are precursors of recombinant IL-2 induced activated killer cells. J Immunol 1985; 134: 802–807.
12. Moretta L, Pende D, Corrani R, Merti A, Bagnesco M, Mirgari MC. T-cell nature of some lymphokine activated killer cells. Frequency analysis of LAK precursors within T-cell populations and clonal analysis of LAK effector cells. Eur J Immunol 1986; 16:1623–1625.
13. Tilden AB, Itoh K, Balch CM. Human lymphokine activated killer cells: Identification of two types of effector cells. J Immunol 1987; 138:1068–1073.
14. Rosenstein M, Ettinghausen SE, Rosenberg SA. Extravasation of intravascular fluid mediated by the systemic administration of recombinant interleukin-2. J Immunol 1986; 137: 1735–1742.
15. Damle NK, Doyle LV, Bender JR, Bradley EC. Interleukin-2 activated human lymphocytes exhibit enhanced adhesion to normal vascular endothelial cells and cause their lysis. J Immunol 1987; 138:1779–1785.
16. Katasek D, Ochoa AC, Vercellotti GM, Bach FH, Jacob HS. LAK cell mediated endothelial injury: A mechanism for capillary leak syndrome in patients treated with LAK cells and IL-2. Clin Res 1987; 35:660A.
17. Parkinson DR, Syndman DR, Weisfuse B, Werner B, Graham D, et al. Hepatitis A (HAV) infection occurring following IL-2/LAK cell therapy. Proc ASCO 1987; 6:234.
18. Martin LK, Sentinella K. An effective dressing technique for large lumen central venous catheters. Crit Care Nurse 1988; (Submitted).
19. West WH, Teuer KW, Yannelli JR, Marshall GD, Orr DW, et al. Constant-infusion recombinant interleukin-2 in adoptive immunotherapy of advanced cancer. N Engl J Med 1987; 316:898–905.

5

Lymphokine–Activated Killer Lymphocytes: Biotherapeutics Clinical Trials

WILLIAM H. WEST
Biotherapeutics, Inc., Memphis, Tennessee

The search for cellular immune mechanisms active against human neoplasia has taken a tortuous course. An early claim that peripheral blood lymphocytes from cancer patients selectively kill tumor cells in microcytotoxicity assays (1) was followed by the recognition that lymphocytes from normal individuals possess the same cytotoxic reactivity (2). Thus, came the description of a natural killer (NK) cell. Efforts to identify tumor-specific T cells by culture of lymphocytes in high-dose interleukin-2 (IL-2) led to the recognition of a broadly reactive killer cell clearly distinct from cytotoxic T cells (CTL) (3). Thus came the description of the lymphokine-activated killer cell (LAK cell). More recently, experience with long-term culture of lymphocytes and the expansion of lymphocytes directly from tumor tissue has led to the recognition that tumor-reactive T cells do exist in humans (4). The pendulum has therefore swung full circle, from an early hypothesis that tumor-specific cytotoxic cells exist in the blood streams of cancer patients to recognition that NK and LAK cells can be identified in normal controls as well as in cancer patients, and now back to the observation that tumor-reactive T cells do exist in some cancer patients, identifiable by their propensity to home to tumor tissue.

Although research continues into the relative contribution of NK, LAK, and CTL in immunosurveillance, effort is also underway to enlist these various killer cells in the treatment of established cancer (4–6). Each of these cytotoxic cell subpopulations can be triggered to proliferate and activate by exposure to IL-2. The production of large quantities of IL-2 by recombinant technique permitted the development of cancer treatment protocols involving the administration of high doses of IL-2 and the adoptive transfer of large numbers of IL-2–activated cells. Peripheral blood mononuclear cells can be harvested in bulk by leukapheresis, activated by IL-2 in highly automated systems, and reinfused with IL-IL-2 in a process termed *adoptive cellular* therapy (ACT). In addition to the generation of large numbers of cytotoxic cells with antitumor potential, adoptive cellular therapy also results in the triggering of a "lymphokine cascade" with secretion of multiple lymphokines with potential antitumor effects (7).

As a laboratory-based company specializing in the development of biological therapy for cancer, Biotherapeutics, Inc. has made a major commitment to the production of activated cells suitable for adoptive immunotherapy. A companion chapter in this text outlines our contribution to standardization of cell activation in the laboratory. It is gratifying to recognize that the culture of billions of lymphocytes for cancer treatment has become a routine laboratory exercise. In this chapter we will focus on our clinical experience with adoptive cellular therapy. We will review the evolution of our clinical protocol involving interleukin-2 and LAK cell therapy. We will also describe ongoing derivative studies that we hope will contribute to continued progress in this dynamic area.

EVOLUTION OF THE LAK CELL PROTOCOL

Studies of adoptive cellular therapy in humans owe their origin to two scientific developments in the 1980s. First, was the successful gene cloning of IL-2 with production of an abundant supply of recombinant lymphokine for preclinical and clinical testing (8). This facilitated rapid completion of Phase I studies, providing early information about toxicities and dose effects of rIL-2 in humans. The second key development was the successful modeling of cellular therapy in experimental animal models at the National Cancer Institute (NCI; 9,10). These animal studies provided crucial information about the effects of IL-2 alone, LAK cells alone, and the synergistic antineoplastic effects of LAK cells in combination with IL-2. These preclinical studies provided the foundation for design of cellular therapy studies in patients with cancer.

In the design of our clinical trial of adoptive cellular therapy we chose a continuous infusion schedule for rIL-2 (6). Given the short half-life of rIL-2 in human serum, we questioned the bolus dosing of rIL-2 in an 8-hr schedule in NCI trials. Such bolus dosing would be expected to result in sharp peak-and-trough serum levels of IL-2, and we hypothesized that cytotoxic cells might be more efficiently stimulated in vivo by constant levels of rIL-2. Animal experiments suggested that antitumor effects might be more pronounced with continuous, rather than intermittent, rIL-2 administration (10). We also suspected that a continuous-infusion schedule might facilitate the management of clinical toxicities. The administration of maximum-tolerated doses of rIL-2 in a 15-min bolus infusion can be expected to trigger prolonged hypotension in a significant proportion of patients (11). The administration of biologically equivalent doses of IL-2 by continuous infusion, in contrast, might produce hypotension of more gradual onset and permit dose titration on a timely basis. Given the capillary leak syndrome associated with high-dose of rIL-2, we also hoped that a continuous infusion schedule, with its attendant ability to titrate dose, would enable us to minimize intravenous fluids and, thereby, reduce the risk of cardiopulmonary morbidity.

The appropriate dose and duration of rIL-2 administered by continuous infusion was not well known in the early days of our studies. In a small pilot trial of patients treated at 1, 2, or 3 × 10^6 units/m^2 per day for 5 days (IL-2 kindly provided by the Cetus Corporation, Emeryville, California), we were impressed with a threshold phenomenon. Patients treated below the 3 × 10^6 units/m^2 per day tolerated rIL-2 well, but developed only a modest lymphocytosis. Patients treated at 3 × 10^6 units/m^2 per day, in contrast, began to manifest dose-limiting toxic reations in the form of hypotension, oliguria, and a rising creatinine level. These patients also demonstrated a dramatic increase in the number of lymphocytes reentering the bloodstream at completion of the IL-2 infusion. Table 1 presents the rebound lymphocytosis seen 36 hr after a 5-day priming infusion of rIL-2 in patients receiving 1 or 2 × 10^6 units/m^2 a day versus that of patients receiving a full 3 × 10^6 units/m^2 a day. With these data in mind, we have placed great emphasis on the height of the rebound lymphocytosis 36 hr after the IL-2–priming cycle as an important index of

Table 1 Impact of IL-2 Priming Dose on Lymphocytes

IL-2 dose (× 10^6 μ/m^2/day)	Rebound lymphocyte count[a]	
1 or 2	2405 (± 1209)	
		$p < .001$
3	6810 (± 5159)	

[a]36 hours after the initial 5-day infusion IL-2.

biological effect. As discussed later, patients who fail to develop a lymphocytosis in excess of 6000 cells/mm^2 have been unlikely to manifest tumor response in our studies of rIL-2–LAK.

The development of oliguria and a rising creatinine level has prohibited administration of continuous infusion rIL-2 at 3 × 10^6/m^2 a day for more than 5 days. The rest interval of 5 days that follows this priming cycle of rIL-2 has been dictated by the time required to capture the migration of lymphocytes by leukapheresis and to activate peripheral blood mononuclear cells in vitro. During periods of leukapheresis, we have focused on preparation for subsequent therapy, with transfusion as necessary to restore a hematocrit of 30%, with diuresis to bring body weight back to baseline levels, and with attention to exercise and caloric intake. Elevated creatinine levels have invariably returned to baseline levels during this period, with only an occasional patient requiring additional days of rest for normalization of renal function.

After 5 days of rest and leukapheresis, patients have received a second period of continuously infused IL-2 to accompany infusion of activated cells. We have typically reinfused cells over a 3-day period, pooling cells from the last 2 days of leukapheresis for the third day of reinfusion. Although our goal was originally to "drive" reinfused cells with 5 days of rIL-2, we have subsequently shortened the driving cycle of rIL-2 to 4 days to avoid excessive fatigue, hypotension, or elevated creatinine levels.

Perhaps the most significant modification of our protocol has involved the timing of additional treatment after completion of the first period of cellular therapy. In our initial protocol, patients proceeded immediately with additional leukapheresis. These patients then received a second period of cell reinfusion, with yet another "driving" cycle of rIL-2. Patients completing this protocol were hospitalized for 25 days with only two 5-day periods of rest separating three separate infusions of rIL-2. It was apparent that these patients were excessively fatigued by the completion of the 25-day program. Recovery to pretreatment performance was delayed up to 3 to 4 weeks, and there was no possibility for repeat treatment within 2 months. In addition, we were dissuaded from this 25-day protocol by changes in the toxicity profile during the third infusion of rIL-2. Patients actually manifested less biological effect and less toxicity during this third cycle of rIL-2. It was apparent that levels of rIL-2 receptor had accumulated on cells or in the serum to create an rIL-2 "sink," or alternatively, that immunoregulatory circuits were blunting the effect of additional rIL-2. Despite a dose escalation scheme that permitted increases of rIL-2 to as high as 7×10^6 units/m^2 a day, it was apparent that there was diminishing effect in this third cycle of rIL-2. The decision was made to interrupt treatment after the initial reinfusion of activated cells (i.e., after 15 days of treatment) both to avoid excessive fatigue and to permit resolution of possible inhibitory factors. Since January 1987, patients have completed one 15-day period of therapy and have then been discharged home for 3 weeks rest. Upon return for additional treatment, patients have repeated an identical 15-day period of cellular therapy. Under this scheme patients receive more IL-2 than in the original 25-day protocol, that is, four infusions of rIL-2, rather than three, but with less cumulative fatigue. Patients have generally been able to be discharged from the hospital within 48 hr of completing these 15-day cycles of treatment and have returned to baseline performance within 10 days. More importantly, the recycling of treatment after periods of rest appears to have affected the quality of tumor response. Some of the most impressive periods of tumor regression have occurred with resumption of IL-2 after 3-weeks rest. All patients destined to achieve either partial or complete regression during this second cycle of adoptive cellular therapy, and several patients who were "stable" after initial treatment, demonstrated major tumor response during retreatment.

By shortening periods of cellular therapy to 15 days and, thereby, avoiding excessive fatigue, we have also been able to add monthly infusions of maintenance rIL-2 for stable or responding patients. After completing two 15-day periods of cellular therapy, patients have then rested again for 3 weeks before proceeding with 4-day infusions of rIL-2 every month for 4 months. Table 1 summarizes the evolution of our cellular therapy protocol. Figure 1 outlines the final version of our LAK cell protocol combining two

Table 2 Major Biotherapeutics LAK Protocol Modifications 1986-1987

1. January 1986–Escalation of the dose of rIL-2 to 3 × 10^6 units/m^2 per day. Identification of a threshold phenomenon manifested by a dose-dependent increase in rebound lymphocytosis.
2. September 1986–Abbreviation of treatment from 25 days to 15 days.
3. September 1986–Recycling of cellular therapy after 3 weeks rest.
4. January 1987–Addition of monthly maintenance rIL-2.

periods of cellular therapy with four cycles of maintenance rIL-2 over a 6-month period.

CLINICAL RESULTS

It is now well recognized that adoptive cellular therapy can result in regression of metastatic cancer of various histologic types and at various sites of disease. As outlined in a summary of patients treated in our Memphis, Tennessee, program, we have witnessed 2 complete responses (CR), 20 partial responses (PR), and 4 minor responses (MR) among 98 patients we could evaluate. Eleven additional patients developed medical problems or disease progression that precluded the completion of cell reinfusion and were considered inevaluable for response. We have confirmed that kidney cancer and melanoma can respond to treatment with rIL-2 and rIL-2–activated mononuclear cells (Fig. 2). We have also observed responses in lung, ovarian, thyroid, and parotid cancers, and in lymphoma. Sites of response have included skin, lymph node, lung, mediastinum, pleura, liver, spleen, and bone. In addition to responses in these sites with intravenous administration of LAK cells, we have witnessed regression of colon cancer metastatic to liver after infusion of LAK cells into the hepatic artery, or mesothelioma involving the peritoneum following intraperitoneal infusion of LAK cells, and of lung cancer involving the pleural after intrapleural infusion of LAK cells. This cumulative ex-

Table 3 Summary of Responses

Histology	CR rate	PR rate	MR rate	Progression rate	CR + PR/ total
Melanoma	0	8	2	12	8/22
Renal	2	4	2	11	6/19
Colon	0	0	0	22	0/22
Lung	0	2	0	7	2/9
Ovary	0	1	0	3	1/4
Breast	0	0	1	4	0/5
Hodgkin's	0	1	0	1	1/2
Non-Hodgkin's lymphoma	0	1	0	2	1/3
Mesothelioma	0	1	0	1	1/2
Parotid	0	1	0	1	1/2
Thyroid	0	1	0	0	1/1
Others[a]	0	0	0	7	0/7
Total	2	20	5	71	22/98

[a]One patient each with carcinoid, hepatoma, gastric, nasopharyngeal, cervix, uterus sarcoma.

perience suggests that a broad array of cancers are sensitive to killing by rIL-2-activated cells or their lymphokine products.

We have attempted to identify clinical indications that correlate with the likelihood of response to adoptive cellular therapy. Almost all responding patients in our studies have been fully ambulatory. As one might anticipate, partially bedridden patients are less able to tolerate the cumulative toxic effects of high-dose rIL-2 and are unlikely to achieve a major tumor response. In addition, almost all patients who have achieved partial or complete response have demonstrated a rebound lymphocytosis exceeding 6000 cells/mm^2 after the priming cycle of rIL-2. Although the height of the rebound lymphocytosis may be simply a reflection of performance status or tumor burden, it has proved a useful indicator of the biological effect of rIL-2. Patients with a disappointing lymphocytosis have generally failed to demonstrate meaningful regression. Although it has been difficult to quantitate the effect of tumor burden on the likelihood of tumor response, it has been unusual to witness a response in the presence of bulky tumor masses exceeding 10 × 10 cm. This may at least partially explain our observation that all patients with kidney cancer achieving partial or complete response have undergone prior nephrectomy.

Whereas there has been abundant evidence of biological effect of adoptive cellular therapy in the form of transient minor or mixed responses, it has been more difficult to achieve durable complete or near-complete responses. With kidney cancer, our ability to recycle cellular therapy in a predictable fashion may have influenced the durability and quality of response. Of 19 patients with kidney cancer who entered into our study, there have been two with complete and four with partial responses. Two of these responding patients were treated with a single 25-day period of cellular therapy before protocol modifications permitted recycling of treatment. Both of these patients manifested disease progression within 2 months. In contrast, the remaining four patients with complete or partial responses have undergone recycling of therapy. Response duration for these patients have equalled 8, 10, 12+, and 18+ months. Confirmation that recycling of therapy prolongs the duration of response will require analysis of many patients, but we are encouraged by these preliminary observations.

With melanoma, the effect of recycling of therapy is not as well defined. As with renal cancer, it has been possible to detect transient biological effect in over 50% of patients with melanoma. The incidence of true partial response, in contrast, has been more modest. No patient has achieved complete response, and the longest duration of partial response has been 5 months. This failure to achieve a durable complete response in melanoma has provided impetus to protocol modifications described in the following discussion.

The observation that two patients with melanoma who had a partial response from adoptive cellular therapy subsequently achieved complete response after treatment with dacarbazine (DTIC) chemotherapy provided the rationale for a separate study in melanoma of alternating cellular therapy and dacarbazine chemotherapy. Dacarbazine was administered during the 3-week rest periods between cycles of cellular therapy. Thus far in this ongoing study, it is unclear that this strategy has resulted in a substantial increase in the major response rate or in the achievement of complete response.

Although our experience with other cancers is too limited to permit conclusions on the efficacy of cellular therapy in any single type, lung cancer and lymphoma would appear to be appropriate targets for an in depth study. We have documented complete regression of a hilar mass in a patient with adenocarcinoma of the lung. We have also observed near-complete resolution of adenopathy and splenomegaly in a patient with nodular mixed lymphoma. The latter patient remains clinically stable 20 months after treatment.

FUTURE DIRECTIONS

Although we can point to progress in both the laboratory and the clinic in the application of cytotoxic cells in cancer treatment, numerous questions remain regarding this new approach to the biotherapy of cancer. The appropriate dose and schedule of rIL-2 remains unclear. The delivery of rIL-2 by continuous infusion at a dosage of 3×10^6 units/m^2 day is clinically manageable in a standard oncology unit. One might argue, however, that the depth and durability of response might be favorably affected by intensification of treatment. The levels of lymphocytosis achieved in our continuous infusion studies appear clearly superior to those reported in bolus-dose studies (12). We take this as evidence that the steady serum levels of rIL-2 achieved with continuous infusion are associated with optimal lymphoproliferation. On the other hand, it is theoretically possible that the peak serum levels of IL-2 achieved by periodic bolus doses might contribute to an antitumor effect, perhaps by triggering additional secretion of the lymphokines, such as tumor necrosis factor (TNF) or interferon-γ. To test this hypothesis and to permit titration of rIL-2 dose intensity for each individual patient, we have designed a hybrid schedule for IL-2 administration that we hope will preserve the advantages of both continuous infusion and bolus-dose administration. Patients will continue to receive rIL-2 by continuous infusion at 3×10^6 units/m^2 a day. In addition, they will receive daily infusion of rIL-2 as a 15-min bolus, at 1 to 3×10^6 units/m^2 a day, titrated according to patient tolerance. Additional pilot trials will also address issues of immunoregulation by including cyclophosphamide as part of treatment plans. Cyclophosphamide has been synergistic with rIL-2 in studies of cellular therapy involving tumor reactive T cells (13). Although animal studies have failed to demonstrate an advantage for cyclophosphamide in association with LAK cells, it is probable that at least a component of response to IL-2–LAK in humans reflects stimulation of underlying T-cell responses. Pretreatment of patients with cyclophosphamide may enhance the ability of IL-2–LAK to act on tumor infiltrating T cells in vivo.

In addition to ongoing efforts to identify optimal schedules and doses of rIL-2 and to explore the effect of pretreatment with cyclophosphamide, we are also exploring combination lymphokine therapy in Phase I trials. Various lymphokines, including TNF, interferon, and IL-2, are reported to have synergistic antitumor effects (14). By identifying active combinations of lymphokines in separate pilot trials, it should be possible to develop a strong rationale for combination lymphokine therapy in conjunction with adoptively transferred cells.

On a more fundamental level, IL-2–LAK cell therapy may be inherently limited by its failure to achieve a sustained immune response to tumor-associated antigens. The LAK cells do not possess the fine specificity of T cells. They also require exposure to high levels of rIL-2 for maximum tumor killing and become quiescent in the absence of rIL-2. Tumor-reactive T cells, in contrast, might confer a state of immunity after adoptive transfer, and T-cell responses may be less dependent on the exogenous administration of high-dose rIL-2. Adoptive transfer of tumor-infiltrating T cells has been demonstrated to be more effective than LAK cell treatment in an experimental animal model (13). With these considerations in mind, we have initiated studies of adoptive cellular therapy involving T cells cultured directly from tumor tissue. Although our studies of cellular therapy involving tumor-derived T cells have only recently begun, we have observed a major response in a patient with melanoma who had previously failed treatment with IL-2–LAK.

The availability of recombinant human lymphokines and the refinement of cell-culturing techniques will lead to an explosion of trials involving cellular therapy of cancer in humans. Combined studies of LAK cells, T cells, immunoregulatory chemotherapy drugs, and multiple lymphokines will evolve in the near future. There is a high likelihood that some form of cellular therapy will become a standard component of cancer care. Although it is highly gratifying to see this rapid evolution of new technologies, it is equally exciting to witness the acquisition of major insight into host–tumor interaction. We can finally identify, isolate, expand, and transfer cytotoxic cells in great numbers. We should be able to accurately define the role of these cells in tumor resistance. We hope we can also determine the degree to which manipulation of those cells will result in meaningful benefit for the tumor-bearing host.

REFERENCES

1. Hellstrom I, Hellstrom K E, Sjogren H O, Warner G A. Demonstration of cell-mediated immunity to human neoplasms of various histological types. Int J Cancer 1971; 7:1–16.
2. Herberman R D, Ortaldo J. Natural killer cells: Their role in defenses against disease. Science 1981; 214:24–36.
3. Grimm E A, Mazumder A, Zhang H A, Rosenberg S A. Lymphokine-activated killer cell phenomenon. Lysis of natural-killer resistant fresh solid tumor cells by interleukin-2–

activated autologous human peripheral blood lymphocytes. J Exp Med 1982; 155:1823–1828.

4. Topalian S L, Solomon D, Avis E P, Chang A E, Freerkser D L, et al. Immunotherapy of patients with advanced cancer using tumor-infiltrating lymphocytes and recombinant interleukin-2: A pilot study. J Clin Oncol 1988; 6:839–853.
5. Rosenberg S A. Lymphokine activated killer cells. A new approach to immunotherapy of cancer. J Natl Cancer Inst 1985; 75:595–603.
6. West W H, Tauer K W, Yannelli J R, Marshall G D, Orr D W, et al. Constant infusion recombinant interleukin-2 in adoptive immunotherapy of advanced cancer. N Engl J Med 1987; 316:898–905.
7. Lotze M T, Matory Y L, Ettinghausen S E, et al. In vivo administration of purified human interleukin-2. II. Half-life, immunological effects, and expansion of peripheral lymphoid cells in vivo with recombinant interleukin-2. J Immunol 1985; 135:1865–1875.
8. Taneguchi T, Matsui H, Fujita T, et al. Structure and expression of a cloned CDNA for human interleukin-2. Nature 1983; 302:305–310.
9. Mule J J, Shu S, Rosenberg S A. The anti-tumor efficacy of lymphokine-activated killer cells and recombinant interleukin-2 in vivo. J Immunol 1985; 135:646–652.
10. Ettinghausen S E, Rosenberg S A. Immunotherapy of murine sarcomas using lymphokine activated killer cells. Optimization of the schedule and route of administration of recombinant interleukin-2. Cancer Res 1986; 46:2784–2792.
11. Rosenberg S A, Lotze M T, Muul L M, Chang A E, Avis F P, et al. A progress report on the treatment of 157 patients with advanced cancer using lymphokine-activated killer cells and interleukin-2 or high-dose interleukin-2 alone. N Engl J Med 1987; 316:889–897.
12. Fisher R I, Coltman C A, Doroshow J H, Rayner H A, et al. Metastatic renal cell cancer treated with interleukin-2 and lymphokine-activated killer cells. A Phase II clinical trial. Ann Intern Med 1988; 108:518–523.

13. Rosenberg S A, Spiess P, Lafreniere R. A new approach to the adoptive immunotherapy of cancer with tumor infiltrating lymphocytes. Science 1986; 233:1318–1321.
14. Winkelhake J L, Stampfl S, Zimmerman R J. Synergistic effects of combination therapy with human recombinant interleukin-2 and tumor necrosis factor in murine tumor models. Cancer Res 1987; 47:3948–3953.

6

Lymphokine-Activated Killer Lymphocytes: LAK and Interleukin-2 in the Treatment of Malignancies of the Central Nervous System

J. BOB BLACKLOCK and ELIZABETH A. GRIMM
M. D. Anderson Hospital, The University of Texas System Cancer Center, Houston, Texas

Neoplastic processes occurring in the central nervous system (CNS) have remained a very difficult problem in oncology. Two manifestations of CNS malignancy have been the subjects of recent investigations utilizing lymphokine-activated killer cells (LAK): primary astrocytic tumors, and leptomeningeal spread of tumor from primary CNS processes or from distant malignancies. These disease states have dismal clinical prognoses with poor response rates to conventional therapy (1,2).

The major primary CNS malignancy in the adult population comprises malignant astrocytomas occurring in the cerebral hemispheres. The most common form of this tumor is glioblastoma multiforme followed by the less common anaplastic astrocytomas. Unfortunately, these tumors are frequently grouped together under the term *malignant glioma*. Careful grading of these tumors by specific histological criteria is important, because crucial differences are evident from the history of each tumor type. These differences are reflected in the clinical outcome of these diseases, which is also correlated to the age of the patient within the respective disease category. In a review of 1440 malignant astrocytic tumors in the cerebral hemispheres of adults, Burger and his associates found a median survival of approximately 24 months with anaplastic astrocytoma (3). All of these patients were treated on Phase III protocols in which they received surgery, radiation therapy, and chemotherapy.

Because patient survival is not the major issue in the current discussion, the general term *glioma* will be used to discuss both. With rare exception, gliomas do not spread outside of the CNS, and their growth is usually limited to local or regional expansion within the cerebral hemispheres. Seeding of the cerebral spinal fluid (CSF) may occur occasionally. This progression by local growth has encouraged the idea that focal therapy with stimulated immune cells is worthy of investigation. The stimulation of peripheral lymphocytes with interleukin-2 (IL-2), with the resultant transformation to LAK cells with its attendant tumoricidal states was first reported by Grimm and her associates in 1982 (4,5). These LAK cells were found to be very active in their ability to lyse autochthonous natural killer (NK)-resistant, fresh uncultured melanoma, sarcoma, and adenocarcinoma. Glioma cells have proved to be equally sensitive in vitro. Jacobs et al. demonstrated that fresh and cultured human glioma cells are sensitive to in vitro lysis by autologous LAK cells (6). The ability to stimulate lymphocytes from brain tumor patients was not inhibited by a history of having received radiation therapy, chemotherapy, or concurrent administration of glucocorticoids. A Phase I study of intracavitary and intralesional application of autologous LAK cells was conducted by Jacobs and his associates (7). Patients received either IL-2 (up to 10^6 units) or LAK cells (up to 10^{10} cells), or both, by direct intracerebral injection around the operative resection cavity. No toxicity attributable to IL-2 or LAK cells was encountered in these patients. Efficacy of therapy cannot be evaluated from the results of this Phase I study. Interest in this technique has been stimulated and several centers are investigating the local applications of LAK cells in the treatment of glioma.

SPECIAL PROBLEMS ASSOCIATED WITH INTERFACING LAK CELL THERAPY WITH CENTRAL NERVOUS SYSTEM MALIGNANCIES

A tumor in the CNS presents special difficulties in applying conventional therapeutic modalities because of its sensitive location, the constraints of the blood–brain barrier, the sensitivity of the brain to edema with potentially disastrous results, and the sensitivity of the blood supply to the brain. The application of large fluid volumes and shifts of fluid into the edematous brain, when associated with a tumor, present considerable risk and may prevent the useful application of systemic IL-2 and LAK. The blood-brain barrier in a glioma is disrupted, which is evidenced by the simple technique of iodine contrast enhancement for computed tomographic scanning. Unfortunately, the neoplastic invasion extends beyond the area of disrupted blood–brain barrier, and thus the absence of the blood–brain barrier in the tumor represents only a portion of the viable neoplastic process (8).

The LAK-mediated tumor cell lysis requires cell-to-cell contact with relatively high effector/target ratio. Currently the delivery of LAK cells to tumor cells is viewed by us as the major problem in the use of this type of therapy against any tumor, and it is the most formidable task in the treatment of gliomas. As yet, glioma cells do not appear to be immunogenic, as they have not been found to express unique antigens recognized by the cellular immune system (9–11). Thus, systemic administration of cytotoxic lymphocytes through the vascular system may not result in significant numbers of the activated cells in the tumor because antigen-specific homing has not yet been achieved. Intra-arterial application of chemotherapy has been applied to brain tumors with the attendant increase in pharmacological advantage. Intra-arterial application of LAK cells would increase the available pool of cells to the tumor bed, but the risk of embolization of small vessels in the brain is an obstacle when considering this type of delivery. Although such plugging of small vessels in some organ systems is acceptable and could even be considered therapeutically advantageous, it could result in unacceptable damage to brain tissue. Because of these inherent difficulties in treating brain tumors with this type of therapy, attention has been turned to local application.

Direct application of large numbers of LAK cells in conjunction with local or systemic stimulation with IL-2 may circumvent some of the obstacles to delivering potentially therapeutic doses of LAK cells. There is no evidence that LAK cells are migratory within tissue beds, and gliomas are known to be infiltrative in the substance of the brain; studies addressing the migration of LAK are currently in progress by us. Intracavitary or intralesional application of even large numbers of LAK cells into operative sites in gliomas or into the substance of gliomas is unlikely to result in a distribution of cells in such a way that meaningful numbers of LAK cells can come into contact with a substantial portion of the tumor cells. These encumbrances to the application of current LAK cell therapy to this focal disease in the brain continue to be vexing problems in applying LAK or other cytotoxic therapeutic modality to brain tumors.

Leptomeningeal spread of primary neoplastic processes in the CNS, or spread of non-CNS malignancies, into spinal fluid pathways results in a generally therapeutically refractory burden of leptomeningeal disease (LMD). This occurrence does present an interesting avenue for the pursuit of using LAK cells in the CNS. The craniospinal axis in adults is bathed with approximately 150 ml of CSF. The nervous system is a relatively privileged compartment in the human because of the blood brain and blood CSF barrier. These barriers have been used to the advantage of the therapist in treating LMD with direct instillation of conventional chemotherapeutic agents into CSF (12). Leptomeningeal tumor produces sheets of tumor in the subarachnoid space attached to the leptomeninges, nerve roots, spinal cord, brain stem, cranial nerves, cerebellum, and cerebrum, as well as free-floating malignant cells in the CSF itself. This relatively closed compartment with circulating CSF may provide an ideal medium for LAK cells to exhibit oncolytic activity. We have reported that LAK are oncolytic in autochthonous CSF (13). In the milieu of the CSF, LAK cells infused into the ventricular system have a relatively high potential for obtaining cell-to-cell contact in numbers that may realistically be oncolytic. Additionally, IL-2 may have an extended half-life compared with that in the systemic circulation, thus allowing for continued stimulation of the LAK cells (14). Administration of LAK cells or IL-2, or both, directly into the spinal fluid is practical, and it can be carried out through a standard Ommaya reservoir. Recently, Shimizu et al. have reported the treatment of several patients with LMD who were treated with direct instillation of LAK cells or IL-2 through an Ommaya reservoir or a ventriculoperitoneal shunt (15). Although clinical efficacy is unclear, these patients did not suffer untoward side effects from the treatment. If LAK cells have a therapeutic role in neoplasia involving CSF pathways, then further investigation into the use of LAK in the adjunctive treatment of the craniospinal axis is in order. Prophylactic craniospinal irradiation in children could potentially be supplanted by adjunctive use of CSF-borne LAK cells in diseases such as medulloblastoma.

CONCLUSION

The CNS offers distinct challenges and opportunities for the investigation of adoptive cellular immunotherapy. The CNS allows for direct instillation of stimulated lymphocytes either into the tumor bed or into the CSF. The limitations are related to the delicacy of the organs involved, the critical blood supply, the blood–brain barrier, and the difficulty in interpreting survival data. The possibility of using LAK cells in the treatment of leptomeningeal spread of certain tumors occurring in the central nervous system is very appealing. Further investigation into these modalities is certainly warranted and is now ongoing at a number of centers.

REFERENCES

1. Walker M D, Green S B, Byar D R, Alexander E, Batzdorf U, Brooke W H, Hunt W E, MacCarty C S, Mahaley M S, Mealey J, Owens G, Ransohoff J, Robertson J T, Shipiro W R, Smith

K R, Wilson C B, Strike T A. Randomized comparisons of radiotherapy and nitrosourea for the treatment of malignant glioma after surgery. N Engl J Med 1980;303:1323–1329.
2. Wasserstrom W R, Glass K P, Posner J B. Diagnosis and treatment of leptomeningeal metastases from solid tumors: Experience with 90 patients. Cancer 1982;49:759–772.
3. Burger P C, Vogel F S, Green S B, Strike T A. Glioblastoma multiforme and anaplastic astrocytoma, pathologic criteria and prognostic implications. Cancer 1985;56:1106–1111.
4. Grimm E A, Ramsey K M, Mazumder A, Wilson D J, Djeu J, Rosenberg S A. Lymphokine-activated killer cell phenomenon II: Precursor phenotype is serologically distinct from peripheral T lymphocytes, memory cytotoxic thymus-derived lymphocytes and natural killer cells. J Exp Med 1983;157: 884–897.
5. Grimm E A, Mazumder A, Zhang H Z, Rosenberg S A. Lymphokine-activated killer cell phenomenon: Lysis of natural killer-resistant fresh solid tumor cells by interleukin-2 activated autologous human peripheral blood lymphocytes. J Exp Med 1982;155:1823–1841.
6. Jacobs S K, Wilson D J, Kornblith P L, Grimm E A. In vitro killing of human glioblastoma by interleukin-2 activated autologous lymphocytes. J Neurosurg 1986;64:114–117.
7. Jacobs S K, Wilson D J, Kornblith P L, Grimm E A. Interleukin-2 and autologous lymphokine-activated killer cells in the treatment of malignant glioma. J Neurosurg 1986; 64: 743–749.
8. Blacklock J B, Oldfield E H, DiChiro G, Tran D, Theodore W, Wright D C, Larson S M. Effect of barbiturate coma on glucose utilization in normal brain versus gliomas. Position emission tomography studies. J Neurosurg 1987;67:71–75.
9. Gately M K, Glaser M, McCarron R M, Dick S J, Dick S J, Dick M D, Mettatal R W, Kornblith P L. Mechanisms by which human gliomas may escape cellular immune attack. Acta Neurochir 1982;64:175–197.
10. Grimm E A, Jacobs S K, Lanza L, Roth J A, Wilson D J. Is there a role for IL-2 activated cytotoxic lymphocytes (LAK)

in cancer therapy? In Kripke M, Frost P (eds): Immunology and Cancer. Austin, University of Texas Press, 1986:209–220.

11. Kumar S, Taylor G, Steward J K, Waghe M A, Morris-Jones P. Cell-mediated immunity and blocking factors in patients with tumors of the central nervous system. Int J Cancer 1973; 12: 194–205.
12. Theodore W H, Gendelman S. Meningeal carcinomatosis. Arch Neurol 1981; 38:696–699.
13. George R E, Loudon W G, Moser R P, Bruner J M, Steck P A, Grimm E A. In vitro cytolysis of primitive neuroectodermal tumors of the posterior fossa (medulloblastoma) by lymphokine-activated killer cells. J Neurosurg 1988; 69:403–409.
14. Lotze M T, Matory Y L, Ettinghausen S E, Raynor A A, Sharrow S O, Seipp C A, Custer M C, Rosenberg S A. In vivo administration of purified human interleukin-2. II. Half-life, immunologic effects and expansion of peripheral lymphoid cells in vivo with recombinant IL-2. J Immunol 1985; 135: 2865–2875.
15. Shimizu K, Okamoto Y, Miyao Y, Yamada M, Ushio Y, Hayakawa T, Ikeda H, Mogami H. Adoptive immunotherapy of human meningeal gliomatosis and carcinomatosis with LAK cells and recombinant interleukin-2. J Neurosurg 1987; 66:519–521.

7

Activated Killer Monocytes: Preclinical Model Systems

HIROTO SHINOMIYA and MIZUHO SHINOMIYA
National Cancer Institute, Frederick, Maryland

G. W. STEVENSON
Northwestern University Children's Hospital, Chicago, Chicago, Illinois

HENRY C. STEVENSON
National Cancer Institute, Bethesda, Maryland

Since Metchnikoff discovered the importance of mononuclear phagocytes (monocytes and macrophages) in the host defense systems of lower animals, much attention has been paid to finding applications for this cell type in mammals, including humans. Among other functions, the role of monocytes in host defense against malignant cells has attracted increasing attention over the last decade. There has been growing in vitro and in vivo experimental evidence to indicate that macrophages are capable of destroying tumor cells. This seems to be reasonable, because malignant cells are also known to exist in invertebrate animals in which mononuclear phagocytes constitute the only immunological host defense mechanism. It is now felt that activated killer monocytes (or macrophages; AKM) seem to be able to selectively kill a wide variety of tumor cells by somehow distinguishing them from phenotypically normal cells (1-3). This simple function may enable us to apply these effector cells to the treatment of patients with cancer.

Before 1980, most research studies focusing on AKM-mediated tumor cell destruction were performed in vitro using murine peritoneal macrophages (1,4–10). A brief summary of the conclusion of these studies is as follows:

1. Normal mouse macrophages are not tumoricidal under usual physiological conditions. To become AKM, they have to undergo some specific functional modifications in vivo with activators such as Bacillus Calmette Guerin (BCG) and *Corynebacterium parvum*, or in vitro with other macrophage-activating factors (MAF), such as polyinosinic-polycytidylic acid (poly I:C), bacterial lipopolysaccharides, or the interferons.
2. The tumor cell cytolytic activities of murine AKM are rather short-lived (< 48 hr) if they are isolated from an MAF-rich environment. The tumoricidal potential of murine macrophages can be maintained by continuous stimulation with MAF.
3. Murine AKM are able to kill diverse tumor cell types even though the tumor cells may be resistant to killing by other modalities (such as anticancer drugs); AKM do not harm normal host cells.
4. The mechanisms by which AKM preferentially recognize tumor cells are not well defined; in some cases, the elaboration of secretory factors, such as tumor necrosis factor, can selectively lyse some tumor cells. In murine experimental systems, a poorly characterized, cell-to-cell contact mechanism seems to play a central role in AKM-mediated tumor cell death.

Given the previous findings from the mouse models, considerable research on the cytotoxic capabilities of human AKM has been performed, chiefly with peripheral blood monocytes (because of the safety and convenience of their collection). Moreover, in 1983, the first AKM-based adoptive immunotherapy treatment of cancer patients was performed as a result of the unique capabilities possessed by human AKM (11,12). Therefore, there is increasing need of laboratory methods for evaluating the antitumor functions of these human effector cells. In this chapter we will summarize previous studies in this field and propose optimal in vitro and animal models for assessing the functions of human AKM.

ISOLATION OF HUMAN MONOCYTES/MACROPHAGES

Mononuclear phagocytes (monocytes and macrophages) are known to distribute almost everywhere in the body as summarized in Table 1. Studies performed on the antitumor activities of macrophage precursors (such as monoblasts or promonocytes) are rare. It is also difficult to obtain human tissue macrophages on a regular basis. Almost all of the human AKM research has been performed with peripheral blood monocytes because of their availability. Although monocytes are not exactly the same cells as macrophages, current evidence suggests that peripheral blood monocytes are differentiated enough to exert antitumor activities in vitro (13–36). An additional special source for human AKM is found in the tumor specimens themselves. Special interest has been paid to tumor-associated macrophages (TAM) and their potential activity in limiting cancer growth and spread (37–39).

Table 1 Sources of Human Monocytes/Macrophages for Preclinical Assays

Peripheral blood monocytes
freshly isolated monocytes
in vitro cultured monocytes
Tissue macrophages
peritoneal macrophages
alveolar macrophages
Kupffer cells
bone marrow macrophages
Tumor-associated macrophages (TAM)

The isolation methods for mononuclear phagocytes are more meticulous than those of any other kinds of human immunological cell types, because these ameba-like cells are very sensitive to stimulation from the external environment. It should be noted that the functions and characteristics of purified monocytes/macrophages may be substantially modified by the isolation procedures used or by contaminants found in the final cellular product.

Representative methods for isolating peripheral blood monocytes are shown in Table 2. Most previous research on human blood monocytes has been performed with monocytes isolated by classic one-step Hypaque–Ficoll density-gradient separation, followed by plastic adherence (Table 2, I.A). The purity of monocytes obtained by this technique largely depends on washing procedures performed after adherence on a plastic surface. This washing method may be somewhat subjective, and it was reported that some types of cells, such as NK cells or platelets, cannot always be eliminated by this procedure. A more rigorous method is a two-step density-gradient separation (Hypaque-Ficoll plus Percoll), combined with final adherence separation (Table 2, IB); however, in addition to the limitations of working with adhered monocytes, these procedures suffer from the small number of cells they are able to purify (40).

In contrast to these adherence-based methods, the method of counter-current centrifugal elutriation (CCE) makes it possible to isolate a large number of blood monocytes ($\sim 10^9$/2-hr run) in a suspension state (Table 2, II; 41). The CCE-purified monocytes seem to reflect the original characteristics they possessed while circulating in suspension in blood vessels; in contrast, evidence has accumulated to indicate that adherence purification procedures activate certain functions of monocytes.

Specific techniques are necessary for isolating tissue macrophages from each organ they occupy in a body; these include bronchial lavage for alveolar macrophages, peritoneal lavage for peritoneal macrophages, and liver dissociation to obtain Kupffer cells. However, once removed from the original tissue or organ, macrophages can be purified by the same adherence methods

Table 2 Methods for Isolating Human Peripheral Blood Monocytes

I. Adherence-based Methods

A. One-Step Density Gradient Separation

Human Peripheral Blood
↓
Hypaque-Ficoll Density Gradient
↓
Mononuclear Leukocytes (Majority are Lymphocytes)
↓
Adherence on a Charged Glass or Plastic Surface
↓
Wash 5-6 Times (to Remove Lymphocytes)
↓
Adherent Monocytes

B. Two-Step Density Gradient Separations

Human Peripheral Blood
↓
Hypaque-Ficoll Density Gradient
↓
Mononuclear Leukocytes
↓
Percoll Continuous Gradient (Enriched for Monocytes)
↓
Adherence on a Plastic Surface
↓
Wash 2-3 Times
↓
Adherent Monocytes

II. Nonadherent Methods

Human Peripheral Blood
↓
(Cytapheresis)
↓
Hypaque-Ficoll Density Gradients
↓
Elutriator (Fractions 11, 12, 13 Contain Highly Purified Monocytes)
↓
Monocytes in Suspension

shown in Table 2, I.A and I.B (42). Interestingly, CCE had also recently been applied to purification of tissue macrophages (hepatic and alveolar). Human tumors contain variable percentages of tumor-associated macrophages (TAM) (39,43–46): After operative resection of tumors, TAM can be isolated by appropriate dissociation of tumor tissue by physical means, enzyme digestion, and other techniques including CCE (47,48).

IN VITRO ASSESSMENT OF THE ANTITUMOR ACTIVITIES OF HUMAN AKM

One of the prominent features of macrophages is their functional versatility. Although we will focus on the particular functions of macrophages related to their antitumor activities, the ability of AKM to kill tumor cells is not thought to be attributable to any single function of these effector cells. Therefore, there are a number of different methods to evaluate the spectrum of functions that contribute to the antitumor activities of AKM.

Tumor Cell Growth Inhibition Assay

Growth inhibition, as determined by growth inhibition assays (GIA), of tumor growth in vitro by AKM has been demonstrated by reduced numbers of target cells, reduced target cell clonability, inhibition of tumor cell mitoses, and the reduced ability of target cells to incorporate ^{3}H- or ^{125}I-labeled nucleotides into DNA (Table 3). Measurement of radiolabeled nucleotide incorporation into DNA of target cells has advantages over counting cells by visual methods by virtue of the rapidity of these assays and the capability to treat objectively large numbers of samples (49,50).

There are many potential problems with the interpretation of GIA data; for example, thymidine produced by AKM is an inhibitor of tumor cell ^{3}H incorporation, and contact inhibition of target cell growth (a form of cytostasis) occurs commonly

Table 3 Procedure for Growth Inhibition Assay (GIA)

Target cells:	MBL-2 cell line [mouse lymphoma (NK resistant) cell line]
Effector cells:	Human monocytes/macrophages
Effector/target (E/T) ratio:	3:1, 10:1, 30:1, 90:1
Target label:	Methyl-[^{3}H] thymidine
Culture medium:	RPMI-1640 + 20% fetal calf serum
Culture plate:	60-well flat-bottom Terasaki plates
Assay procedure:	10 μl of medium in each well of plates (varying numbers of cells depending on E/T ratio) + 10 μl of effector cell suspension + 10 μl of MBL-2 cell suspension (2 × 10^5/ml) + 1 μl of methyl [^{3}H] thymidine (0.25 μCi/μl) ↓ Incubate for 24 hr at 37°C in a 5% CO_2 The culture plates are harvested and the radioactivity incorporated in MBL-2 cells is determined by counting the disks in a liquid scintillation counter.

among cells cultured in vitro. Moreover, the GIA cannot distinguish tumor cell cytolysis from cytostasis. Cytostasis can be detected earlier than cytolysis; however, it remains to be determined if cytostasis must appear in advance of cytolysis, or even if it shares common cellular functions with cytolysis.

Tumor Cell Cytotoxicity Assays

Cytotoxic activities are more aggressive functions of AKM. This in vitro direct antitumor activity is thought to be the most important function of these effector cells and has the most potential correlation with the application of human AKM to the clinical treatment of cancer. Cytotoxic activities of AKM have been examined by a variety of methods; visual quantitation, measurement of released radioactivity from labeled target cells, and measurement of radioactivity in residual adherent tumor targets after rigorous washing. Additional assay variables include the kinds of target tumor cells, activator substances (BRM), culture vessels, and culture medium employed in each assay. These variables have caused some impediments to progress in this field of monocyte study according to a recent consensus workshop (51). Therefore, there is a need to have a standardized assay of cytotoxic activities of AKM in which a standard medium, culture vessel, BRM activator, and set of target tumor cells are employed. We will summarize the methods currently used and introduce a novel AKM cytotoxicity assay recently developed by our group.

Rapid Assays

Recently, European groups developed a rapid assay method (6–18 hr) for measuring the cytolytic function of unstimulated human blood monocytes against the murine tumor cell line WEHI-164 pretreated with actinomycin D (52–54). According to these authors, this system has several advantages, including the lack of sensitivity of the target cells to natural killer (NK) cells, and the short assay time allows study of monocytes before extensive maturation. Subsequent studies have demonstrated that a tumor necrosis factor (TNF)-like monokine released from cultured monocytes is the effector molecule in this type of target cell lysis (10,55). The major problem with this assay seems to be that drug-pretreated xenogeneic tumor cells become too sensitive to human monocytes and to endotoxins contaminating the culture medium (10,55). Therefore, this method may overestimate the cytotoxic activity of unactivated monocytes. Furthermore, the relevance of this in vitro phenomenon to human clinical situations is difficult to visualize.

Long-Term Assays

When intact tumor human target cells and unstimulated human blood monocytes are coincubated, it has been shown that the unactivated effector cells require a longer incubation time to show tumoricidal activity than after activation with BRMs. Two long-term assay systems (24-72 hr) are commonly employed to monitor the cytotoxic capabilities of human monocytes against human tumor targets. The method of Le and co-workers employs CCE-purified monocytes with [^{131}I]IUDR-labeled adherent human tumor cell line targets. Polystyrene culture vessels, fetal calf serum-containing medium, and a variety of interferon stimulators are employed; tumor cell death is measured as a function of the amount of [^{131}]IUDR released into the culture medium at 48 to 72 hr (56). The method of Kleinerman et al. employs human monocytes purified by Hypaque–Ficoll–Percoll gradients plus adherence cocultured with [^{131}]IUDR-labeled adherent human tumor cell line targets. Polystyrene culture vessels, serum-containing medium and a variety of BRM are employed; tumor cell death is measured as a function of the amount of IUDR-labeled viable tumor cells that remain adherent to the culture plate at 48 to 72 hr (18). Methodologic and clinical relevance issues aside, substantial problems persist for existing human monocyte-mediated tumor cell cytotoxicity assay systems according to a recent consensus workshop (51). To try to resolve these problems we have developed a novel AKM cytotoxicity assay using human monocytes, negatively selected in a suspension state by CCE, cultured in chemically-defined serum-free medium (free from endotoxin or lot-to-lot variation) and newly developed culture plates made of nonadhesive materials (Teflon or polypropylene) to which monocytes do not adhere (Table 4). The results obtained thus far have shown that our new AKM cytotoxicity assay method requires lower effector/target cell ratios and lower concentrations of BRM to obtain the same level of target lysis, compared with the foregoing conventional adherence-based assay system (as shown in Fig. 1). This serum-free suspension culture assay system should provide great advantages in evaluating AKM activity without confusion provided by unknown and uncontrollable exogenous events. Most importantly, this assay system is clinically relevant; one is able to activate and manipulate AKM in this assay system in a fashion identical with the methods employed

Table 4 Long-Term (48–72 hr) Human AKM Cytotoxicity Assays

Effector cells	Adherence-purified human monocytes	Adherence-purified human monocytes	Suspension-purified human monocytes
Target cells	Variety of adherent human cell lines	Human adherent A375 melanoma	Variety of nonadherent human cell lines
Target label	$[^{125}I]$ IUDR	$[^{125}I]$ IUDR	^{111}In oxine
Culture medium	Serum-containing	Serum-containing	Serum-free
Culture plate	96-well flat adhesive material	96-well flat adhesive material	96-well round nonadhesive material
Culture time	72 hr	72 hr	48–72 hr
Assay method	Measure radioactivity release into the culture supernatant	Measure radioactivity remaining on the plate after washing	Measure radioactivity released into the culture supernatant
Ref.	Le et al. (56)	Kleinerman et al. (18)	Shinomiya H, et al, 1988 (unpublished)

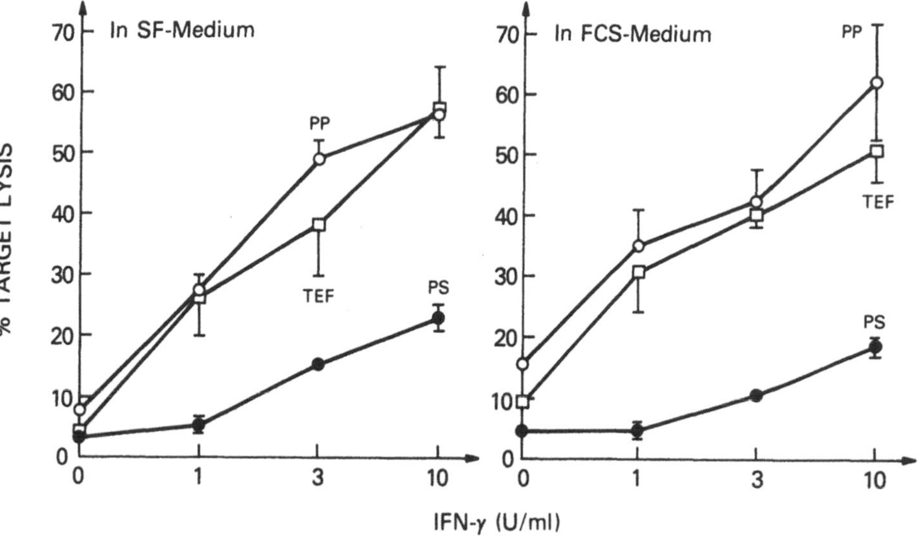

Figure 1 Differences in measured human AKM cytotoxic activity against ^{111}In-labeled colon cancer targets (HT-29) over 3 days when cultured with serum or under serum-free (SF) conditions; also shown is the effect of adherence (polystyrene) and nonadherent (NA) conditions (polypropylene) on measured cytotoxic function. Human recombinant interferon-γ (IFN-γ) is the monocyte/macrophage-activating factor (MAF).

in the performance of AKM adoptive immunotherapy of cancer patients.

SECRETION OF ANTITUMOR SUBSTANCES FROM MONOCYTES/MACROPHAGES

Macrophages secrete over 60 distinct BRM substances. Several of them are known to have antitumor effects (cytostatic or cytotoxic); these include peroxide, superoxide, and tumor necrosis factor-β. Because these BRM may be responsible for certain aspects of AKM-mediated tumor cell cytostasis or cytotoxicity, it is useful to monitor these factors in assessing the functional status of AKM. Moreover, from a practical viewpoint, biochemical measurement of these factors is far easier to perform than cellular biological measurement of AKM-mediated cytostasis or cytotoxicity; assays of reactive oxygen intermediates (32,57-60) and TNF seem to be the most popular of these types of studies (61-63).

CURRENT STATUS OF AKM CYTOTOXICITY ASSAYS

Monocytes and macrophages do not exert their optimal cytostatic, cytotoxic, or secretory activities unless they are activated with appropriate stimuli. A series of BRM that can enhance the tumoricidal activity of monocytes/macrophages have collectively been termed macrophage-activating factor(s) (MAF). Because this nomenclature was determined from a phenomenological standpoint, MAF include a wide variety of substances ranging from physiological substances (such as bacterial products or cytokines) to synthetic chemicals (Table 5). The MAF can work individually, and the combination of MAF may work synergistically (in some cases) or, in others, actually suppress one another (10,18,20,21, 27-29,40,64-75). From a clinical standpoint, it seems desirable to be able to determine the individual capability of each MAF or their combinations to generate AKM in each individuals; thereby allowing for formulation of the best MAF and dosage for each cancer patient.

Our results indicate that normal donor-derived AKMs tested in the serum-free suspension-culture system are able to kill a wide range of human suspension-cultured tumor targets. For the first time, we have observed excellent killing at low effector/target (E/T) ratios (as low as 1:1) with reproducible increases in cytotoxicity seen with increasing E/T ratios. Baseline (unactivated) cytotoxicity is very low; excellent killing is induced by coincubation of the monocytes with a variety of BRM (and combinations thereof; see Fig. 1). Virtually every human tumor tested is killed by human AKM within a 48-hr timespan; no normal cell types are killed by human AKM. Endotoxin is not required as an AKM activator.

Table 5 Representative Human Monocyte/ Macrophage Activating Factor(s) (MAF)

1. Endogenous BRM
   IL-1 = Interleukin-1
   IL-2 = Interleukin-2
   BSF1 = B-Cell stimulating factor I
   IFN-γ = Gamma Interferon
2. Exogenous biologicals
   LPS = Lipopolysaccharide
   MDP = Muramyldipeptide
   Poly I:C = Polyinosinic:polycytidilic acid
3. Combinations of 1, 2, or 1 and 2
   IFN-γ + LPS
   IFN-γ + MDP
   Others

MONOCYTE/MACROPHAGE FUNCTIONS IN HUMAN CANCER PATIENTS

As shown earlier, monocytes/macrophages have a variety of measurable antitumor activities. However, even though *normal* monocyte/macrophages from *healthy* donors can attack tumor cells in vitro, this does not immediately mean that monocytes/ macrophages from *cancer* patients have the same function. Because malignant cells somehow succeed in becoming established tumors in spite of host defense mechanisms, it has been surmised that at least parts of the immune systems of cancer patients do not work effectively or that malignant cells somehow suppress the function of certain components of the immune system. Taking this into consideration, it is important to assess antitumor activities of monocytes/macrophages from patients with cancer. Those monocytes/macrophage studies that have been performed to date can be categorized into two groups: those performed with

Table 6 Examples of Laboratory Assessments of Cancer Patient Monocyte Cytotoxic Activity

Patients studied	Leukocyte source	Assay performed	Compared to normal control values	Reference
Breast, colon, head and neck, and melanoma	Blood monocytes	Cytotoxicity + ADCC cytotoxicity	Variable	76
Renal cancer	Blood monocytes	Cytotoxicity	↓	77
Ovarian cancer	Blood monocytes	Cytotoxicity	↓	78
Ovarian cancer	Ascitic macrophages	Cytotoxicity	↓	79
Gastrointestinal tract, lung, breast, and lymphoma	Blood monocytes	ADCC cytotoxicity	↓	80
Breast, lung, and melanoma	Blood monocytes	ADCC cytotoxicity	↓	81
Advanced malignancies-various types	Blood monocytes	Superoxide-release	↓	59
Gastrointestinal cancer	Blood monocytes	Superoxide-release	↓	82
Pediatric malignancies	Blood monocytes	Chemotaxis	↓	83
Advanced malignancy-various types	Blood monocytes	IL-1 production	↓	84
Hepatocellular cancer	Blood monocytes	IL-1 production	↓	85
Gastrointestinal cancer	Blood monocytes	Cytotoxicity and ADCC	↓	86

peripheral blood monocytes and those performed with monocytes/macrophages infiltrating into tumor tissue itself (TAM). Over the past decade, a number of institutions have reported on their experiences with widely differing assay systems for distinct groups of cancer patients (Table 6; 76-86). It is now difficult to make any global statements from these studies, except to suggest the following:

1. The monocyte/macrophage tumoricidal activity in certain groups of cancer patients may be depressed when compared with that of normal persons.
2. A wide range of BRM and biologicals may have the capacity to augment the tumoricidal activity of cancer patients' monocytes and macrophages.

IN VIVO ASSESSMENT OF AKM IN ANIMALS

The design of AKM animal models that may permit prediction of results in humans is an important priority; until recently, little progress has been made in this area. Adoptive transfer of activated peritoneal macrophages from syngeneic mice to mice with pulmonary metastases have tended to give inconsistent results, frequently seeming to promote tumor cell growth in vivo. A recent system developed by Bartholeyns and co-workers uses normal resident peritoneal macrophages (EPM) from BALB/C mice expanded in vitro on a layer of mesothelial cells. When injected intravenously, 1 million EPM were able to cause regression of subcutaneous EMTG sarcoma nodule; this EPM antitumor activity was enhanced by the intraperitoneal injection of lipopolysaccharide following adoptive transfer of the cytotoxic cells (87).

Another animal model approach has been to use nude mice injected with lethal doses of human tumors (such as colon or ovarian carcinoma cell lines). These animals then receive adoptively transferred human AKM (with or without additional BRM); this model has been most extensively tested with peritoneal carcinomatosis models and intraperitoneal delivery mechanisms for the AKM, with or without additional BRM. Results, to date, from studies that used AKM alone have been disappointing; additional BRM (usually IFN-γ) appears to be required to maintain the AKM cytotoxic activity. Coinoculation of tumor cells, plus AKM, plus IFN subcutaneously in nude mice (the Winn assay) is highly effective; results utilizing the intraperitoneal model are somewhat inconsistent. We are currently examining what should prove to be more clinically relevant animal models. By using a larger animal (the miniature swine), we will be able to isolate peripheral blood monocytes in a fashion identical with the clinical trials, including the performance of cytapheresis/CCE; porcine AKM manipulation in vitro will be under serum-free suspension-culture conditions. These animals have a genetic propensity to develop malignant melanoma and should prove ideal for our preclinical animal AKM research.

CONCLUSIONS

Activated killer monocytes (AKM) can be monitored preclinically by a wide range of in vitro assays and animal models. Most suffer from lack of clinical relevance and from irreproducibility. Even the best assays are burdened by their time-consuming nature and the high degree of technical expertise required to execute them. As clinical trials with AKM become more widely carried out, we must strive to identify more effective preclinical model systems whenever possible. Moreover, the entire discipline will profit from consensus conferences (and wet workshops when possible) to finalize rational new standards for these assays.

REFERENCES

1. Hibbs J B Jr. Discrimination between neoplastic and non-neoplastic cells in vitro by activated macrophages. J Natl Cancer Inst 1974;53:1487–1492.
2. Fidler I J, Jessup J M, Gofler W E, Staerkel R, Mazumder, A. Activation of tumoricidal properties in peripheral blood

monocytes of patients with colorectal carcinoma. Cancer Res 1986;46:994-998.
3. Fidler I J, Kleinerman E S. Lymphokine-activated human blood monocytes destroy tumor cells but not normal cells under cocultivation conditions. J Clin Oncol 1984; 2:937-943.
4. Fidler I J, Folger W E. Activation of tumoricidal properties in macrophages by lymphokines encapsulated in liposomes. Lymphokine Res 1982;1:73-77.
5. Hibbs J B Jr, Chapman H A Jr, Weinberg J B. The macrophage as an antineoplastic surveillance cell: Biological perspectives. J Reticuloendothel Soc 1978;24:549-570.
6. Meltzer M S. Tumor cytotoxicity by lymphokine-activated macrophages: Development of macrophage tumoricidal activity requires a sequence of reactions. Lymphokines 1981; 3:319-343.
7. Varesio L. Induction and expression of tumoricidal activity by macrophages. In Dean R T, Jessup W (eds): Mononuclear Phagocytes: Physiology and Pathology. 1985:381-407.
8. Hibbs J B Jr. Heterocytolysis by macrophages activated by bacillus Calmette-Guerin: Lysosome exocytosis into tumor cells. Science 1974;184:468-471.
9. Bucana C, Hoyer L C, Hobbs B, Breesman S, McDaniel M, Hanna M G Jr. Morphological evidence for the translocation of lysosomal organelles from cytotoxic macrophages into the cytoplasm of tumor target cells. Cancer Res 1976;36:4444-4458.
10. Marino P A, Adams D O. Interaction of bacillus Calmette-Guerin-activated macrophages and neoplastic cells in vitro: I. Conditions of binding and its selectivity. Cell Immunol 1980; 54:11-25.
11. Stevenson H C, Foon K A, Sugarbaker P. Ex vivo activated monocytes in adoptive immunotherapy trials in colon cancer patients. Prog Clin Biol Res 1986;211:75-82.
12. Stevenson H C, Miller J P, Beman J A, Ottow R, Abrams P G, Keenan A, Larson S, Woodhouse C, Sugarbaker P. Analysis of the trafficking of purified activated human monocytes fol-

lowing intraperitoneal infusion in colon cancer patients. Cancer Res 1987;47:6100–6103.

13. Fidler I J, Schroit A J. Macrophage recognition of self from nonself: Implications for the interaction of macrophages with neoplastic cells. Symp Fundam Cancer Res 1986; 38:183–207.
14. Fischer D G, Hubbard W J, Koren H S. Tumor cell killing by freshly isolated peripheral blood monocytes. Cell Immunol 1981;58:426–435.
15. Graziano R F, Fanger M W. Human monocyte-mediated cytotoxicity: The use of Ig-bearing hybridomas as target cells to detect trigger molecules on the monocyte cell surface. J Immunol 1987;138:945–950.
16. Itoh K, Platsoucas C D, Balch C M. Monocyte- and natural killer cell-mediated spontaneous cytotoxicity against human noncultured solid tumor cells. Cell Immunol 1987; 108: 495–500.
17. Kleinerman E S, Ceccorulli L M, Bonvini E, Zicht R, Gallin J I. Lysis of tumor cells by human blood monocytes by a mechanism independent of activation of the oxidative burst. Cancer Res 1985;45:2058–2064.
18. Kleinerman E S, Erickson K L, Schroit A J, Fogler W E, Fidler I J. Activation of tumoricidal properties in human blood monocytes by liposome containing lipophilix muramyl tripeptide. Cancer Res 1983;43:2010–2014.
19. Klostergaard J, Foster W A, Hamilton D A, Turpin J, Lopez-Bernstein G. Effector mechanisms of human monocyte-mediated tumor cytotoxicity in vitro: Biochemical, functional, and serological characterization of cytotoxins produced by peripheral blood monocytes isolated by counterflow elutriation. Cancer Res 1986;46:2871–2875.
20. Koff W C, Folger W E, Gutterman J, Fidler I J. Efficient activation of human blood monocytes to a tumoricidal state by liposomes containing human recombinant gamma interferon. Cancer Immunol Immunother 1985;19:85–89.
21. Lopez-Berestein G, Mehta K, Mehta R, Juliano R L, Hersh E M. The activation of human monocytes by liposome-

encapsulated muramyl dipeptide analogues. J Immunol 1983; 130:1500–1502.
22. Mantovani A, Jerrells T R, Dean J H, Herberman R B. Cytolytic and cytostatic activity on tumor cells of circulating human monocytes. Int J Cancer 1979; 23:18–27.
23. Mavier P, Edgington T S. Human monocyte-mediated tumor cytotoxicity. I. Demonstration of an oxygen-dependent myeloperioxidase-independent mechanism. J Immunol 1984; 132:1980–1986.
24. Mukherji B. In vitro assay of spontaneous cytotoxicity by human monocytes and macrophages against tumor cells. J Immunol Methods 1980; 37:233–247.
25. Normann S J, Weiner R. Cytotoxicity of human peripheral blood monocytes. Cell Immunol 1983; 82:413–425.
26. Rinehart J J, Lange P, Gormus B J, Kaplan M E. Human monocyte-induced tumor cell cytotoxicity. Blood 1978; 52:211–220.
27. Saiki I, Sone S, Fogler W E, Kleinerman E S, Lopez-Berestein G, Fidler I J. Synergism between human recombinant gamma-interferon and muramyl dipeptide encapsulated in liposomes for activation of antitumor properties in human blood monocytes. Cancer Res 1985; 45:6188–6193.
28. Schacter B, Kleinhenz M E, Edmonds K, Ellner J J. Spontaneous cytotoxicity of human peripheral blood mononuclear cells for the lymphoblastoid cell line CCRF-CEM: Augmentation by bacterial lipopolysaccharide. Clin Exp Immunol 1981; 46:640–648.
29. Sone S, Lopez-Berestein G, Fidler I J. Potentiation of direct antitumor cytotoxicity and production of tumor cytolytic factor in human blood monocytes by human recombinant interferon-gamma and muramyl dipeptide derivatives. Cancer Immunol Immunother 1986; 21:93–99.
30. Uchida A, Klein E. Natural cytotoxicity of human blood monocytes and natural killer cells and their cytotoxic factors: Discriminating effects of actinomycin D. Int J Cancer 1985; 35:691–699.
31. Uchiyama H, Suzuki T, Oboshi S, Ino H. Cytotoxic activity

of human blood monocytes against cultured breast cancer cells. Gann 1978; 69:259-262.

32. Weiss S J, Slivka A. Monocyte and granulocyte-mediated tumor cell destruction. A role for the hydrogen peroxide-myeloperoxidase-chloride system. J Clin Invest 1982; 69: 255-262.
33. Yanagawa E, Uchida A, Kokoschka E M, Micksche M. Natural cytotoxicity of lymphocytes and monocytes and its augmentation by OK432 in melanoma patients. Cancer Immunol Immunother 1984; 16:131-136.
34. Unsgaard G. Cytotoxicity to tumor cells induced in human monocytes cultured in vitro in the presence of different sera. Acta Pathol Microbiol Scand Sect C Immunol 1979; 87:141-149.
35. Kaplan G. In vitro differentiation of human monocytes. Monocytes cultured on glass are cytotoxic to tumor cells but monocytes cultured on collagen are not. J Exp Med 1983; 157:2061-2072.
36. Mantovani A, Polentarutti N, Peri G, Shavit Z B, Vecchi A, Bolis G, Magnioni C. Cytotoxicity on tumor cells of peripheral blood monocytes and tumor-associated macrophages on patients with ascites ovarian tumors. J Natl Cancer Inst 1980; 64:1307-1315.
37. Eccles S A, Alexander P. Macrophage content of tumors in relation to metastatic spread and host immune reaction. Nature 1974; 250:667-669.
38. Evans R, Alexander P. Cooperation of immune lymphoid cells with macrophages in tumor immunity. Nature 1970; 228:620-622.
39. Svennevig J L, Lovik M, Svaar H. Isolation and characterization of lymphocytes and macrophages from solid, malignant human tumors. Int J Cancer 1979; 23:626-631.
40. Kleinerman E S, Schroit A J, Fogler W E, Fidler I J. Tumoricidal activity of human monocytes activated in vitro by free liposome-encapsulated human lymphokines. J Clin Invest 1983; 72:304-315.
41. Stevenson H C, Fauci A S. Countercurrent centrifugal elutria-

tion of human monocytes. In Hersocowitz H, Holden H, Bellanti M (eds): Manual of Macrophage Methodology. Plenum Press, New York, 1980:75.
42. Loveren H V, Otter W D. Macrophages in solid tumors. I. Immunologically specific effector cells. J Natl Cancer Inst 1974;53:1057-1060.
43. Svennevig J L, Svaar H. Content and distribution of macrophages and lymphocytes in solid malignant human tumors. Int J Cancer 1979;24:754-758.
44. Wood G W, Gollahon K A. Detection and quantitation of macrophage infiltration into primary human tumors with the use of cell-surface markers. J Natl Cancer Inst 1977; 59: 1081-1087.
45. Bottazzi B, Polentarutti N, Balsari A, Boraschi D, Ghezzi P, Salmona M, Mantovani A. Chemotactic activity for mononuclear phagocytes of culture supernatants from murine and human tumor cells: Evidence for a role in the regulation of the macrophage content of neoplastic tissues. Int J Cancer 1983;31:55-63.
46. Bottazzi B, Polentarutti N, Acero R, Balsari A, Boraschi D, Ghezzi P, Salmona M, Mantovani A. Regulation of the macrophage content of neoplasma by chemoattractants. Science 1983;220:210-212.
47. Yasaka T, Wells R J, Mantich N M, Boxer L A, Baehner R L. Enrichment by counterpart centrifugal elutriation of human lymphocytes cytotoxic to human tumor cells. Immunology 1982;46:613-617.
48. Krahenbuhl J L, Remington J S. Inhibition of target cell mitosis as a measure of the cytostatic effects of activated macrophages on tumor target cells. Cancer Res 1977; 37: 3912-3916.
49. Kaplan A M. Cytostasis of tumor and nontumor cells. In Adams D O, Edelson P J, Koren H (eds): Methods for Studying Mononuclear Phagocytes. New York, Academic Press, 1981.
50. Maluish A E, Leavitt R, Sherwin S A, Oldham R K, Herberman R B. Effects of recombinant interferon-alpha on im-

mune function in cancer patients. J Biol Response Mod 1983; 2:470–481.

51. Koren H S, Herberman R B. The current status of human monocyte-mediated cytotoxicity against tumor cells. J Leukocyte Biol 1985; 38:441–445.
52. Colotta F, Peri G, Villa A, Mantovani A. Rapid killing of actinomycin D-treated tumor cells by human mononuclear cells. I. Effectors belong to the monocyte-macrophage lineage. J Immunol 1984; 132:936–944.
53. Colotta F, Bersani L, Lazzarin A, Poli G, Mantovani A. Rapid killing of actinomycin D-treated tumor cells by human monocytes. II. Cytotoxicity is independent of secretion of reactive oxygen intermediates and is suppressed by protease inhibitors. J Immunol 1985; 134:3524–3531.
54. Ziegler-Heitbrock H W, Riethmuller G. A rapid assay for cytotoxicity of unstimulated human monocytes. J Natl Cancer Inst 1984; 72:23–29.
55. Kornbluth R, Edington T S. Tumor necrosis factor production by human monocytes is a regulated event: Induction of TNF-α-mediated cellular cytotoxicity by endotoxin. J Immunol 1986; 137:2585–2591.
56. Le J, Prensky W, Yip Y K, Chang Z, Hoffman T, Stevenson H C, Balazs I, Sadlik J R, Vilcek J. Activation of human monocyte cytotoxicity by natural and recombinant immune interferon. J Immunol 1983; 131:2821–2829.
57. Ernst M, Heberer M, Fischer H. Chemiluminescence measurements of immune cells—a tool in immunobiology and clinical research. J Clin Chem Clin Biochem 1983; 21:555–560.
58. Heberer M, Ernst M, Durig M, Harder F. Chemiluminenscence of granulocytes and monocytes in diluted whole blood samples: A tumor marker? Cancer Detect Prev 1983; 6:273–280.
59. Nakagawara A, Kayashima K, Tamada R, Onoue K, Ikeda K, Inokuchi K. Sensitive and rapid method for determination of superoxide-generating activity of blood monocytes and its use as a probe for monocyte function in cancer patients. Gann 1979; 70:829–833.

60. Nathan C F, Horowitz C R, de la Harpe J, Vadhan-Raj S, Sherwin S A, Oettgen H F, Krown S E. Administration of recombinant interferon gamma to cancer patients enhances monocyte secretion of hydrogen peroxide. Proc Natl Acad Sci USA 1985; 82:8686–8690.
61. Kelker H C, Oppenheim J D, Stone-Wolff D, Henriksen-DeStefano D, Aggarwal B B, Stevenson H C, Vilcek J. Partial physiochemical characterization of tumor necrosis factor produced in human monocytes and its separation from lymphotoxin. Int J Cancer 1985; 36:69–73.
62. Creasey A A, Reynolds M T, Laird W. Cures and potential regression of murine and human tumors by recombinant human tumor necrosis factor. Cancer Res 1986; 46:5687–5692.
63. Tsujimoto M, Ip Y K, Vilcek J. Tumor necrosis factor: Specific binding and internalization in sensitive and resistant cells. Proc Natl Acad Sci USA 1985; 82:7626–7631.
64. Bucana C D, Hoyer L C, Schroit A J, Kleinerman E S, Fidler I J. Ultrastructural studies of the interaction between liposome-activated human blood monocytes and allogeneic tumor cells in vitro. Am J Pathol 1983; 112:101–111.
65. Grabstein K H, Urdal D L, Tushinski R J, Mochizuki D Y, Price D L, Cantrell M A, Gillis S, Conlon P J. Induction of macrophage tumoricidal activity by granulocyte–macrophage colony-stimulating factor. Science 1986; 232:506–508.
66. Hammerstrom J. In vitro influence of endotoxin on human mononuclear phagocyte structure and function. II. Enhancement of the expression of cytostatic and cytolytic activity of normal and lymphokine-activated monocytes. Acta Pathol Microbiol Scand Sect C Immunol 1979; 87:391–399.
67. Jarstrand C, Blomgren H. Influence of bestatin, a new immunomodulator, on various functions of human monocytes. J Clin Lab Immunol 1982; 9:193–198.
68. Onozaki K, Matsushima K, Kleinerman E S, Saito T, Oppenheim J J. Role of interleukin 1 in promoting human monocyte-mediated tumor cytotoxicity. J Immunol 1985; 135: 341–320.
69. Peri G, Polentarutti N, Sessa C, Mangioniu C, Mantovani A.

Tumoricidal activity of macrophages isolated from human ascitic and solid ovarian carcinomas: Augmentation by interferon, lymphokines and endotoxin. Int J Cancer 1981; 28: 143–152.

70. Sone S, Utsugi T, Shirahama T, Ishii K, Matsuura S, Ogawara M. Induction by interferon-alpha of tumoricidal activity of adherent mononuclear cells from human blood: Monocytes as responder and effector cells. J Biol Response Mod 1985; 4:134–140.

71. Sone S, Utsugi T, Tandon P, Ogawara M. A dried preparation of liposomes containing muramyl tripeptide phosphatidylethanolamine as a potent activator of human blood monocytes to the antitumor state. Cancer Immunol Immunother 1986; 22:191–196.

72. Utsugi T, Sone S. Comparative analysis of the priming effect of human interferon-gamma, -alpha, and -beta on synergism with muramyl dipeptide analog for antitumor expression of human blood monocytes. J Immunol 1986; 136:1117–1122.

73. Zahedi K, Mortensen R F. Macrophage tumoricidal activity induced by human C-reactive protein. Cancer Res 1986; 46: 5077–5083.

74. Zhang S R, Salup R R, Urias P E, Twilley T A, Talmadge J E, Herberman R B, Wiltrout R H. Augmentation of NK activity and/or macrophage-mediated cytotoxicity in the liver by biological response modifiers including human recombinant interleukin-2. Cancer Immunol Immunother 1986; 21:19–25.

75. Thurman G B, Maluish A, Rossio J L, Schlick E, Onozaki K, Talmadge J E, Procopio A D, Ortaldo J R, Ruscetti F W, Cannon C B, Iyer S, Stevenson H C, Herberman R B. Comparative evaluation of multiple lymphoid and recombinant human IL-2 preparations. J Biol Response Mod 1986; 5:85–97.

76. Unger S W, Bernhard M I, Pace R C, Wanebo H J. Alterations of monocyte function in neoplastic disease. Surg Forum 1979; 30:142–144.

77. Unsgaad G, Eggen B M, Lamvik J. Depression of monocyte-

mediated cytotoxicity by renal carcinoma and restoration through therapy. J Surg Oncol 1983;22:51-55.
78. Kleinerman E S, Swelling L A, Howser D, Barlock A, Young R C, Decker J M, Bull J, Muchmore A V. Defective monocyte killing in patients with malignancies and restoration of function during chemotherapy. Lancet 1980;22:1102-1105.
79. Peri G, Polentarutti N, Sessa C, Mangioni C, Mantovani A. Tumoricidal activity of macrophages isolated from human ascitic and solid ovarian carcinomas: Augmentation by interferon, lymphokines, and endotoxin. Int J Cancer 1981;28: 143-152.
80. Unger S W, Bernhard M I, Pace R C, Eanebo H J. Monocyte dysfunction in human cancer. Cancer 1983;51:669-674.
81. Bernhard M I, Pace R C, Unger S W, Wanebo H J. Monocyte-mediated antibody-dependent cell-mediated cytotoxicity and spontaneous cytotoxicity in normals and cancer patients as assayed by human erythrocyte lysis. Cancer Res 1983;43: 4504-4510.
82. Nakagawara A, Ikeda K, Inokuchi K, Kumashiro R, Tamada R. Deficient superoxide-generating activity and its activation of blood monocytes in cancer patients. Cancer Lett 1984; 22:157-162.
83. Tono-oka T, Nakayama M, Ohkawa M, Takeda T, Matsumoto S. Impaired leukocyte mobility and production of monocyte-derived granulotactic factor in pediatric malignant disease during chemotherapy. Tohoku J Exp Med 1981; 134:301-310.
84. Santos L B, Yamada F T, Scheinberg M A. Monocyte and lymphocyte interaction in patients with advanced cancer. Evidence for deficient IL-1 production. Cancer 1985; 56: 1553-1558.
85. Herman J, Kew M C, Rabson A R. Defective interleukin-1 production by monocytes from patients with malignant disease. Interferon increases IL-1 production. Cancer Immunol Immunother 1984;16:182-185.
86. Weiner L M, Steplewski Z, Koprowski H, Litwin S, Comis

R L. Divergent dose-related effects of gamma interferon therapy on in vitro antibody-dependent cellular and non-specific cytotoxicity by human peripheral blood monocytes. Cancer Res 1988;48:1042-1046.

87. Bartholeyns J, Lombard Y, Poindron P. Immunotherapy of murine sarcoma by adoptive transfer of resident peritoneal macrophages proliferating in culture. Anticancer Res 1988; (In press).

8

Activated Killer Monocytes: Use in Clinical Trials

LEOCADIO V. LACERNA
National Cancer Institute, Frederick, Maryland

PAUL H. SUGARBAKER
Emory University Medical Center, Atlanta, Georgia

HENRY C. STEVENSON
National Cancer Institute, Bethesda, Maryland

Blood monocytes and their differentiated tissue counterparts, the macrophages, have been of scientific interest since these cells were first described over 95 years ago by Metchnikoff (1). During the past three decades, numerous immunological effector functions of this cell type have been characterized, including antigen processing (2), accessory cell functions (3), suppressor cell functions (4), antibody-dependent cellular cytotoxicity (5), production of immunoregulatory biological response modifiers (BRM) (6), and secretion of critical components of the complement system (7). Since 1975, however, experimental systems have been developed that have helped define the role of mononuclear phagocytes in regulating tumor growth and the formation of metastases. Human blood monocytes have an ability to recognize and kill tumor targets in vitro that is independent of antibody (6–8); this function may be augmented by such agents as interferon-γ (IFN-γ) (9), muramyl dipeptide (10), and macrophage colony-stimulating factor (M-CSF) (11). Numerous in vitro and preclinical model systems have been

Table 1 Stepwise Approach to the Clinical Testing of the AKM Adoptive Cellular Immunotherapy (ACI)

Testing level		Regional	Systemic
Phase I:	Feasibility/toxicity testing	Completed	Initiated
Phase II:	Optimal immunostimulation and AKM targeting to tumor sites/assessment of efficacy	Initiated	Proposed
Phase III:	Randomized controlled trial between optimal AKM therapy and established conventional therapy	Proposed	Proposed

developed in recent years to monitor the tumoricidal function(s) of human monocytes; progress and controversy in these research arenas are summarized in Chapter 5. The demonstration of human blood monocyte tumoricidal activity and mechanisms for modulating this phenomenon has presented the alluring possibility that human monocyte cytotoxic function could be employed as an effective clinical modality; this chapter will summarize adoptive cellular immunotherapy (ACI) applications of this preclinical observation to cancer patients. We have initiated the first ACI therapy utilizing activated killer monocytes (AKM) to treat patients with cancer (Table 1), which has been termed the AKM protocol (12); this chapter will also summarize our experience in the development of this novel immunotherapeutic approach.

IMPEDIMENTS TO THE DEVELOPMENT OF AKM CANCER THERAPIES

The development of a new cancer ACI protocol utilizing AKM has been hampered by numerous technical and theoretical impediments, including (a) the lack of a methodology for supplying a large number of purified autologous cells that can be employed clinically; (b) the need to identify clinical-grade monocyte-activating substances (FDA approval required); and (c) the lack of AKM-compatible equipment, media, and technology that would provide for the suspension culture of these cells in the absence of antibiotics, animal sera, and other potentially sensitizing agents. To date, most of the obstacles to the manipulations of AKM for clinical purposes have been overcome. We have developed techniques for isolating highly purified autologous blood monocytes in large numbers by combining two blood component separation techniques, cytapheresis (Fenwal Laboratories, Deerfield, Illinois; 3) and countercurrent centrifugal elutriation (CCE; Beckman Laboratories, Palo Alto, California; 14). With these two techniques, it is possible to obtain up to 10^9 human blood monocytes with a purity of 90% or greater by a negative-selection process that allows the cells to remain in suspension (15); these cells are sterile and devoid of antibiotics and toxins. In addition, suspension culture techniques have been devised that allow the maintenance of these normally adherent cells in a serum-free medium (16). Clinical-grade monocyte-activating substances are available; we have demonstrated that both natural and recombinant human IFN-γ preparations are excellent boosters of monocyte antitumor activity against colon cancer cell lines at concentrations as low as 2 units/ml (8). Our in vitro and preclinical analyses of AKM have been extended to other malignancies as well (as summarized in Chap. 7). Further details on the in vitro handling of AKM are reviewed in Chapter 12.

THE AKM PROTOCOL: REGIONAL ACI APPROACH

The AKM protocol is a focused effort to deliver cytotoxic blood monocytes to the site of tumor burden in patients with cancer in an attempt to eradicate otherwise incurable disease. We have recently used IFN-γ–activated AKM to treat metastatic cancer patients in both regional and systemic ACI modalities. In our regional ACI approach, we have adoptively transferred these tumoricidal cells directly into the patient's peritoneal cavity for widespread intraperitoneal malignancies; this is known as intraperitoneal AKM therapy. Our regional AKM patient trial has focused on peritoneal colorectal carcinomatosis (PCC), a metastatic form of colon cancer that afflicts approximately 10% of patients with recurrent metastatic disease and for which no effective therapy exists (17). Figure 1 summarizes the current regional immunotherapy approach, in which AKM are delivered intraperitoneally to cancer patients. Patients with PCC are referred for staging celiotomy and debulking surgery to render them as disease-free as possible; a Tenckhoff catheter is also inserted into the peritoneal space to provide ready access to this cavity when needed. As soon as the patient is stable in the recovery room, the

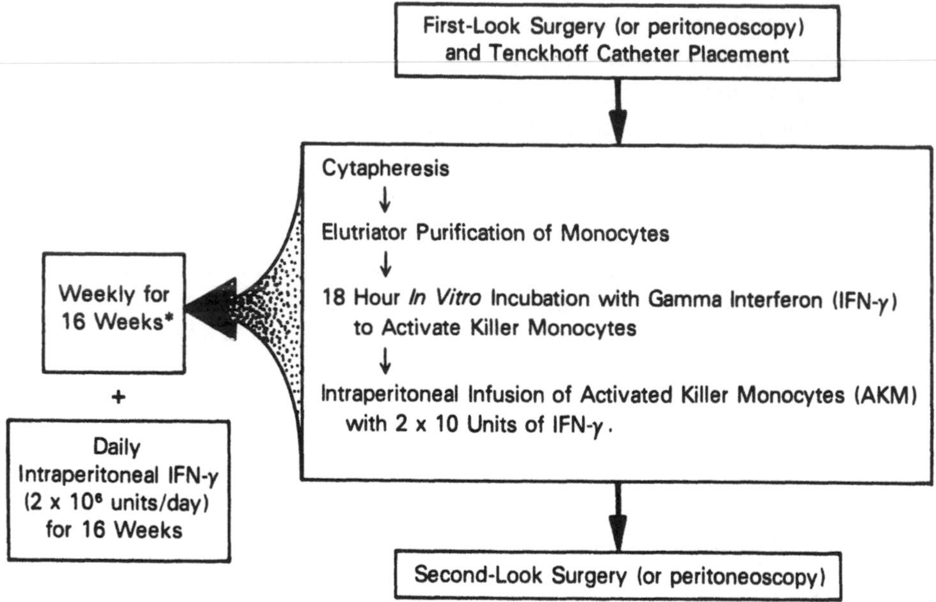

Figure 1 Summary of the present immunotherapeutic approach for intraperitoneal delivery of AKM to colorectal cancer patients.

first infusion of AKM is given intraperitoneally (4–6 $\times$ 10^8 cells). This therapy is given weekly for 16 weeks. Cells are obtained during a 2-hr cytapheresis procedure, after which monocytes are purified by CCE; the purified monocytes are incubated with 1000 units/ml of IFN-γ for 18 hr. A "second-look" staging celiotomy is performed after 16 weeks of regional AKM therapy. This protocol was recently modified to include the intraperitoneal infusion of 2 $\times$ 10^6 units of recombinant IFN-γ on a daily basis for 16 weeks. Midway through the study, the trafficking pattern of IFN-γ–activated AKM is monitored by labeling these cells with ^{111}In before intraperitoneal infusion. Gamma camera imaging has revealed that these cells distribute throughout most of the peritoneal cavity in a relatively homogeneous fashion, with a few patchy areas of decreased uptake, shown at second-look surgery to be due to postoperative adhesions. It appears that activated monocytes infused intraperitoneally remain in the peritoneal space for prolonged periods (perhaps months) and do not traffic outside of the peritoneal space. We were able to obtain interpretable images from patients up to 5 days after intraperitoneal infusion. Furthermore, AKM were infused into a patient 24 hr before a second laparotomy; multiple serosal specimens were obtained and paired samples were stained for nonspecific esterase activity and autoradiography. Examinations of the thin-section esterase-stained slides revealed large numbers of esterase-positive cells on the serosal surface seen infrequently on the peritoneal biopsy specimens obtained at the time of the first laparotomy. The ^{111}In-labeled monocytes could be readily discerned by autoradiographic techniques to be infiltrating into the peritoneal lining. It is probable that the esterase-positive cells that deeply penetrated the peritoneal lining were AKM from previous intraperitoneal infusions that had transformed into macrophages (18).

Toxic reactions from regional AKM therapy consist mainly of low-grade fever and abdominal pain that may occur within 24 hr of the AKM infusion. Approximately 50% of patients have experienced an episode of bacterial peritonitis secondary to chronic implantation of the Tenckhoff catheter; rates of infection are similar to those seen in patients undergoing peritoneal dialysis through a permanent indwelling peritoneal catheter. These infections have responded readily to antibiotics. Moderate leukopenia was noted in some patients secondary to chronic cytapheresis; however, white blood cell counts rarely fell below 3500 cells/mm^3. If mild leukopenia does occur, the frequency of AKM treatments is decreased to once every other week until there is normalization of the peripheral leukocyte count. Table 2 summarizes the complications observed during intraperitoneal AKM therapy. To date, we have treated seven patients on this protocol; six have completed the entire protocol and three have completed maintenance therapy. One patient was removed from the study after eight cycles of treatment because of progressive intraperitoneal disease. Three of the patients have experienced prolonged freedom from local relapse. The longest disease-free interval is 3.5 years. Two patients technically could not be evaluated because they had developed metastatic disease at distant sites. One of these was found to have an elevated level of carcinoembryonic antigen from the eighth cycle on and, therefore, she was not reexplored. She died as a result of a complication that developed during an experimental surgical procedure at another institution. The second technically unevaluable patient developed an 8-mm pulmonary nodule immediately before the second-look celiotomy. This was excised and found to be malignant and, therefore, the patient did not undergo a second abdominal exploration. This patient still manifests systemic disease but is without clinical evidence of intraperitoneal

Table 2 Complications/Toxicity of Intraperitoneal AKM Therapy

Cytapheresis: minimal leukopenia (rarely ≤3500)
Activated killer monocyte (AKM) infusions: local peritoneal irritation, low-grade fever, possible infectious peritonitis
Surgical risks:
staging celiotomies, or
staging peritoneoscopies
Tenckhoff catheter placement and removal

Table 3 Results of 7 PCC Patients treated on the AKM Protocol

Age/sex	Stimulator	Toxicities[a]	Results	Freedom from local relapse
44/F	nIFNγ	Tmax = 101 peritoneal irritation (grade 2)	O-PD	3.5 yrs. posttherapy
41/F	nIFNγ	Afeb.: peritoneal irritation (grade 1): 1 episode-bacterial peritonitis	O-PD	3.0 yrs. posttherapy
52/M	nIFNγ	Afeb.: peritoneal irritation (grade 1) episode-bacterial peritonitis	I-PD, 8 cycles	
42/F	nIFNγ	Afeb.: peritoneal irritation: (grade 1)	I-PD, 16 cycles (no peritoneal disease at autopsy)	
46/M	nIFNγ	Tmax = 100: peritoneal irritation (grade 1) episode-bacterial peritonitis	1 lung metastasis posttherapy, no clinical peritoneal disease	2.5 yrs. posttherapy
31/F	rIFNγ	Tmax = 100: peritoneal irritation (grade 2)	O-PD	2.0 yrs. posttherapy
38/M	rIFNγ	Afeb.: peritoneal irritation (grade 0)	SD, 16 cycles	

Abbreviations: PCC, peritoneal colorectal carcinomatosis; nIFNγ, natural interferon γ; AKM, activated killer monocytes; O-PD, small areas of progressive disease found at second surgery (in areas to which monocytes had limited access)-Pt. rendered surgically disease free; I-PD, peritoneal progressive disease (surgically inoperable); rIFNγ, recombinant interferon γ; SD, stable disease.

[a]Peritoneal irritation grading system: 0 = none; 1 = manageable with oral analagesics; 2 = requiring parenteranarcotics; 3 = requiring intravenous fluids and nasogastric suction.

disease. Table 3 summarizes the results obtained from the regional AKM to date. The patients entering this protocol were felt to have been rendered grossly disease-free by the initial celiotomy procedure. Thus, formal assessments of response rates cannot be measured. On the other hand, remnants of microscopic disease remained postoperatively and the median disease-free interval of such patients treated with surgery alone is 4 months, and nearly all patients have relapsed by 2 years. Therefore, the prolonged disease-free interval seen in three of our patients is an encouraging indication that the AKM therapy may have been useful in eradicating subclinical residual tumor deposits. This possibility is consistent with our current hypotheses for ACI therapy and cancer therapy in general: Attempts at eradication of tumor appear to be most effective with minimal volumes of malignancy.

THE AKM PROTOCOL: SYSTEMIC ACI APPROACH

In an attempt to treat distant metastatic malignant disease, an intravenous AKM immunotherapy protocol was recently established. In this study, peripheral blood monocytes from cancer patients are purified to greater than 90% purity by cytapheresis and CCE. These monocytes are then cultured in suspension in serum-free medium and incubated for 4 hr at 37°C. The cells are labeled with ^{111}In before they are intravenously infused into the patient. If no toxic effects are noted, the patient returns 3 weeks later (to allow time for the ^{111}In emission signals to extinguish) to receive AKM cultured in IFN-γ. These cells are also labeled with ^{111}In, and their clinical effects and trafficking patterns throughout the patient's body are closely monitored. Figure 2 outlines ongoing systemic trafficking studies in patients with distant metastatic cancer. The data from this study have been inconclusive to date, because both unactivated monocytes and AKM appear to traffic principally to the reticuloendothelial system (liver, spleen, and bone marrow) and not to sites of tumor burden. It is theoretically possible that the ^{111}In-labeling procedure may modulate the native trafficking pattern of these cells; however, other assayable

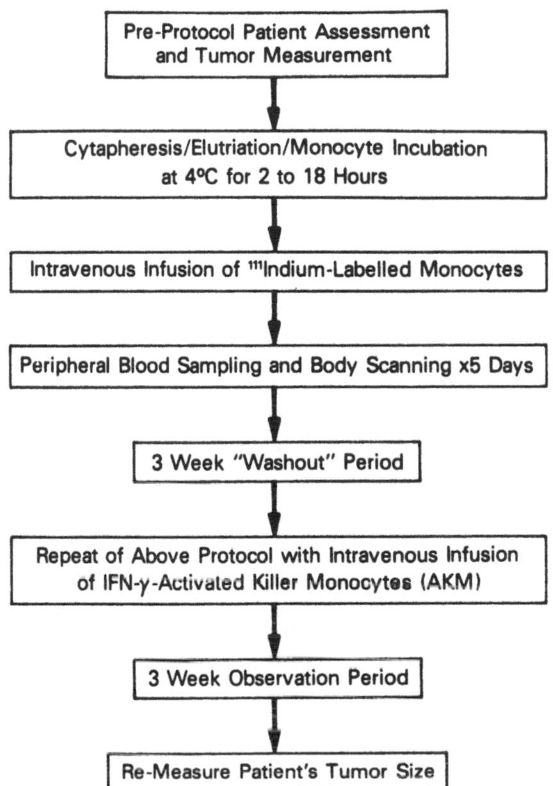

Figure 2 Protocol outline for systemic AKM therapy in patients with distant metastatic cancer.

functions of ^{111}In-labeled monocytes and AKM appear unchanged from baseline. To date, five patients have been treated with this systemic AKM approach with virtually no side effects; all were found to have stable disease at restaging. Strategies to more effectively home AKM to the site of metastatic tumor burden in our patient population are being developed. In addition, it is clear that we have not approached the maximal tolerated dose (MTD) for the systemic administration of AKM; there appears to be substantial opportunities for dose escalation in this ACI approach.

CONCLUSIONS

It is clear that ACI utilizing AKM is a feasible and safe procedure. Although the prolonged freedom from relapse of our three PCC patients treated with the intraperitoneal AKM protocol is encouraging, it is clear that many additional patients must be treated before we can reach any firm conclusions on the efficacy of the treatments for disease control and patient survival. Moreover, attempts to treat patients with bulk disease must be undertaken. Monocyte activators other than IFN-γ must be tested in vitro and in preclinical models with the hope of applying these agents to our regional AKM clinical trials. In addition, we must devise strategies for targeting intravenously infused AKM to sites of metastatic tumor burden in cancer patients requiring systemic therapy. The clinical use of growth factors such as M-CSF which not only boost AKM cytotoxic activity but their numbers as well is a critical next step in the development of more potent AKM therapies. Future generations of AKM therapies and the potential interface of monocyte-based ACI with other forms of ACI (such as LAK and TIL) are reviewed in Chapter 14.

REFERENCES

1. Metchnikoff E. Lectures in the Comparative Pathology of Inflammation. Starling F E (ed). London, Kegan, Paul, Trench, Truber and Co, 1893.
2. Rosenthal A S. Current concepts: Regulation of the immune response role of the macrophage. N Engl J Med 1980; 303: 1153–1156.
3. Stevenson H C. The role of mononuclear phagocytes as accessory cells in lymphocyte responses. In Herscowitz H B, Bellanti J A (eds): The Reticuloendothelial System: A Comprehensive Treatise. New York, Plenum Press, 1983:113–121.
4. Varesio L. Suppressor cells and cancer: Inhibition of immune functions by macrophages. In Friedman H, Herberman R B,

Escober M, Reichard R (eds): The Reticuloendothelial System. New York, Plenum Press, 1983:217-252.
5. Poplack D G, Bonnard G D, Holiman B J, Blaese R M. Monocyte-mediated antibody-dependent cellular cytotoxicity: A clinical test of monocyte function. Blood 1976; 43:809-816.
6. Nathan C F, Murray H W, Cohn Z A. Current concepts: The macrophage as an effector cell. N Engl J Med 1980; 303:622-626.
7. Davies P, Allison A C. Secretion of macrophage enzymes in relation to the pathogenesis of chronic inflammation. In Nelson D S (ed): Immunobiology of the Macrophages. New York, Academic Press, 1976:427-461.
8. Le J, Prensky W, Yip Y K, Chang Z L, Hoffman T, Stevenson H C, Balazs I, Sadlik J R, Vilcek J. Activation of human monocyte cytotoxicity by natural and recombinant immune interferon. J Immunol 1983; 131:2821-2826.
9. Foon K A, Sherwin S A, Abrams P G, Stevenson H C, Holmes P, Maluish A E, Oldham R K, Herberman R B. A Phase I trial of recombinant gamma interferon in patients with cancer. Cancer Immunol Immunother 1985; 20:193-197.
10. Oldham R K, Thurman G B, Talmadge J E, Stevenson H C, Foon K A. Lymphokines, monoclonal antibodies and other biological response modifiers in the treatment of cancer. Cancer 1984; 54:2795-2810.
11. Warren M K, Ralph P. Macrophage growth factor (CSF-1) stimulates human monocyte production of interferon, tumor necrosis factor and colony stimulating activity. J Immunol 1986; 137:2281-2286.
12. Stevenson H C, Foon K A, Sugarbaker P. Ex vivo activated monocytes in adoptive immunotherapy trials in colon cancer patients. Prog Clin Biol Res 1986; 211:75-82.
13. Stevenson H C, Beman J A, Oldham R K. Design of a cytapheresis program for cancer immunology research. Plasma Ther Transplantation Technol 1983; 4:57-63.
14. Stevenson H C. Separation of mononuclear leukocyte subsets by countercurrent centrifugation elutriation. Methods Enzymol 1984; 108:242-249.

15. Akiyama Y, Miller P J, Thurman G B, Neubauer R H, Oliver C, Favilla T, Beman J A, Oldham R K, Stevenson H C. Characterization of a human blood monocyte subset with low peroxidase activity. J Clin Invest 1983;72:1093-1105.
16. Stevenson H C, Schlick E, Griffith R, Chirigos M A, Brown R, Conlon J, Kanapa D J, Oldham R K, Miller P J. Characterization of biological response modifier release by human monocytes cultured in suspension in serum-free media. J Immunol Methods 1983;5:129-146.
17. Sugarbaker P H, MacDonald J A, Gunderson L L. Colorectal cancer. In Devita V T, Hellman S, Rosenberg S A (eds): Cancer, Principles and Practice of Oncology. Philadelphia, J B Lippincott, 1982:643-710.
18. Stevenson H C, Miller J P, Beman J A, Ottow R, Abrams P G, Keenan A, Larson S, Woodhouse C, Sugarbaker P. Analysis of the trafficking of purified activated human monocytes following intraperitoneal infusion in colon cancer patients. Cancer Res 1987;47:6100-6103.

9

Tumor-Infiltrating Lymphocytes from Human Solid Tumors: Antigen-Specific Killer T Lymphocytes of Activated Natural Killer Lymphocytes

THERESA L. WHITESIDE, D. S. HEO, S. TAKAGI, and RONALD B. HERBERMAN
University of Pittsburgh School of Medicine and Pittsburgh Cancer Institute, Pittsburgh, Pennsylvania

The emphasis given to tumor-infiltrating lymphocytes (TIL) in recent studies of adoptive immunotherapy stems from the observation made by Rosenberg and colleagues that TIL were highly effective in causing tumor regression in mice bearing established tumors (1). Murine TIL cultured in vitro in the presence of recombinant interleukin-2 (rIL-2) were claimed to be distinct from lymphokine-activated killer (LAK) cells because of their specificity for autologous tumor targets, better expansion in vitro, and greater antitumor cytolytic potential (1). It has been difficult to experimentally confirm the unique advantage of rIL-2-activated TIL over LAK cells in humans (2). Although IL-2-expanded TIL from three patients with melanoma were reported to show specificity for autologous tumor cells in in vitro cytotoxicity assays (3), this finding has not been as yet reproduced. More recent reports from Rosenberg's laboratory indicate that lytic preference for autologous tumors has been observed only for TIL from human melanomas (4). In most studies, human TIL cultured in high doses of rIL-2 give rise to non–MHC-restricted effectors that do not seem to be qualitatively different from the LAK cells generated from the patient's blood (5). The issue of the specificity of TIL for autologous tumor cells is an important one for the future of adoptive immunotherapy with TIL.

Lymphocytes are often the most prominent cellular component of inflammatory infiltrates into human solid tumors (6). Classic histopathological data have been recently strengthened by immunohistology, which confirmed that T lymphocytes and, in some tumors, B lymphocytes often accumulate in the tumor stroma or invade the tumor parenchyma (2,7,8). The role of these lymphocytic infiltrations in and around solid tumors remains controversial. On the one hand, correlations have been established between the intensity of lymphoid infiltrates and prognosis or survival (8,9). In human tumors for which such correlations exist, (e.g., melanoma, carcinomas of colon, head and neck, breast, lung), the infiltrating lymphocytes are viewed as effectors of local antitumor immunity (10). On the other hand, in some situations, the presence of lymphocytes in tumors could not be related to improved prognosis (e.g., certain melanomas; 11) and is, therefore, considered a result of nonspecific inflammatory interactions. Recent functional studies with freshly isolated and in vitro-cultured TIL from human solid tumors have failed generally to confirm that these tumors contain cytotoxic T lymphocytes (CTL) specific for autologous tumor cells (12–14).

The nature and local functions of TIL in solid human tumors remain unclear, as does the precise relationship between these cells and tumor growth. The reasons for the lack of information about cellular mechanisms that lead to a failure of tumor rejection in humans are varied. Difficulties associated with a recovery of TIL and viable tumor cells in numbers sufficient for functional studies have been considerable. Only advanced tumors removed during therapeutic surgery are available, and only in vitro studies are possible in humans. Still, methods have been developed for the recovery from these tumors of viable TIL and tumor cells (15,16) and have allowed analyses of effector cell functions. Importantly, freshly isolated TIL were found to be poor in vitro effectors of antitumor cytotoxicity. Our own, and other, studies also documented depressed proliferative function of fresh TIL (17,18) as well as their profound unresponsiveness to different activating stimuli including mitogens, alloantigens and TPA (17). The refractory responses of TIL to activating agents and our inability to outgrow tumor-specific T lymphocytes from human TIL preparations in most cases has led to a considerable disappointment, particularly because studies in murine models of tumor growth yielded undisputed evidence for the existence of such T lymphocytes in murine tumors (19,20).

The availability of rIL-2 made it possible to investigate in some detail the characteristics of lymphocytes outgrown from human solid tumors. The studies described here were designed to (a) examine the potential of TIL from several different human tumors for expansion in rIL-2; (b) determine if TIL proliferate better than autologous peripheral blood lymphocytes (A-PBL) in the presence

Table 1 TIL Preparations Obtained from Human Solid Tumors and Used in the Functional Studies Described

Squamous cell carcinomas of the head and neck (SCCHN)	
$n = 29$ (paired TIL and A-PBL)	
primary	26
metastatic	3
Nonsquamous cells cancers of the head and neck	
$n = 4$ (paired TIL and A-PBL)	
primary	3
metastatic	1
Hepatic carcinoma	
$n = 24$ (TIL only)	
primary (hepatocellular CA)	12
metastatic (adeno CA, colon)	12
Ovarian adenocarcinoma	
$n = 9$ (TIL only)	
all primary	

of 1000 units/ml of rIL-2; (c) compare the antitumor effector function of IL-2-expanded TIL and A-PBL; (d) establish the nature of the cell(s) mediating the antitumor activity in IL-2-expanded cultures; and (e) determine if tumor specificity could be demonstrated in IL-2-expanded TIL effector populations. These studies were performed with TIL obtained from three histologically different tumor types as listed in Table 1.

TUMOR-INFILTRATING LYMPHOCYTE EXPANSION IN RECOMBINANT INTERLEUKIN-2

The tumor biopsies or ascites fluids were always obtained fresh from surgery and processed immediately after arrival in the laboratory. The enzymatic digestion and separation of TIL from tumor cells were performed as described by us earlier (7). The TIL preparations recovered from differential density gradients varied in their purity from 60% to 95%, depending on a tumor. The viability of TIL was always > 95%. In contrast, the viability of recovered tumor cells was generally lower, ranging from 75% to 95%. Whenever possible, tumor cells were cryopreserved in 90% pooled AB human serum in a control-rate freezing device to be used in cytotoxicity assays. Fresh TIL, if recovered in sufficiently large numbers, were stained with labeled monoclonal antibodies to the lymphocyte surface antigens before culture and analyzed in a flow cytometer. This enabled us to determine the phenotypic characteristics of cells obtained from enzymatic digests of tumor biopsies. Table 2 shows the results obtained with 17 fresh TIL preparations and A-PBL from patients with primary head and neck cancer. The T lymphocytes were a major component of the TIL preparations, and many of these T cells were in an activated state as judged by the expression of the HLA-DR antigens and receptors for IL-2. The paucity of NK cells ($Leu19^+$, $CD16^+$) in TIL was a consistent feature in these tumors, most of which were advanced (Stages III and IV). A decrease in the proportion of $CD4^+$ cells resulted in the alteration of the CD4/CD8 ratio in TIL: $CD8^+$ cells were relatively more numerous in TIL than in A-PBL preparations

Table 2 Phenotypic Analysis of Fresh TIL and A-PBL from Patients with Head and Neck Cancer[a]

Phenotype	TIL n=17	% Positive cells	A-PBL n=17
CD3+	80 ± 9		70 ± 11
CD2+	84 ± 8		78 ± 18
CD4+	33 ± 13		44 ± 17
CD8+	32 ± 16		30 ± 15
Leu19+	3 ± 2*		10 ± 2
CD16+	4 ± 3*		21 ± 26
CD4:CD8	1		1.5
CD3+Leu19+	2 ± 1.7		5 ± 5
CD3+IL2R+	14 ± 11*		7 ± 6
CD2+HLA-DR+	29 ± 17*		5 ± 4
CD16+/CD8+	1 ± 1		4 ± 4

[a]Data presented as means ± SD. Asterisks indicate differences between the groups. TIL or A-PBL were incubated with labeled monoclonal antibodies to lymphocyte surface antigens purchased from Becton-Dickinson, Sunnyvale, CA, washed in PBS-Na azide buffer and analyzed by two-color flow cytometry in a FACStar instrument.

(see Table 2). The enrichment of TIL in $CD8^+$ cells was observed in half of the preparations and was consistent with our earlier data, which showed that variable proportions of $CD4^+$ and $CD8^+$ cells were present in fresh TIL from different solid tumors (6).

When TIL isolated from solid tumors or obtained from ascites fluids were cultured in the presence of high doses of IL-2 and human serum, they usually proliferated successfully (Table 3). However, TIL from metastatic tumors failed to grow in 31% (5/16) cases as opposed to only 15% (7/47) TIL from primary tumor. Kinetics of TIL expansion in culture were studied for both primary and metastatic tumors. The TIL were delayed in their proliferative responses, with a lag period extending to 30 to 40 days

Table 3 Proliferation of TIL Isolated from Primary and Metastatic Human Solid Tumors in Long-Term Cultures[a]

	Range	
	Expansion-fold	Days in culture
Primary tumors:		
Head and neck carcinoma (n=24)	$2\text{-}350 \times 10^3$	24-91
Hepatocellular carcinoma (n=7)	$87\text{-}39 \times 10^6$	52-188
Ovarian carcinoma (n=9)	8-682	42-91
No expansion (n=7)		
Metastatic tumors:		
Head and neck carcinoma (n=4)	100-1250	45-88
Colon adenocarcinoma into liver (n=7)	$324\text{-}5 \times 10^5$	15-250
No expansion (n=5)		

[a]TIL were isolated from fresh tumor biopsies by enzymatic digestion as described earlier (13), washed and plated at the cell concentration of 0.5×10^6/ml in flat-bottom wells of 48-well Costar plates in a complete RPMI 1640 (Gibco), medium containing 5% (v/v) pooled human AB serum, antibiotics and 1000 U/ml of rIL2 (Cetus, Emeryville, CA). The plates were incubated in a humidified atmosphere of 5% CO_2 in air at 37°C and examined daily. Viable cell counts were performed every other day, and cultures were supplemented with fresh complete medium to maintain the cell concentration at $<2 \times 10^6$/ml. The expansion fold was calculated based on the cell numbers achieved at the end of cultures.

for TIL from metastatic tumors. After the lag period, these TIL often grow rapidly and reached high levels of expansion (e.g., 3.9×10^7 in TIL from metastatic liver tumors). These studies indicated that TIL from most, but not all, human solid tumors expand well in the presence of rIL-2 after an initial period of delay that is variable but always longer for metastatic than for primary tumors. The lag period seen with TIL may be related to immunoinhibitory substances produced by tumors (17,21). Such substances could be responsible for poor proliferative and cytotoxic functions seen with fresh TIL (17,22), and their effects may not be readily reversed even in the presence of high doses of rIL-2.

Table 4 Expansion of TIL and Autologous Peripheral Blood Lymphocytes (A-PBL) in Long-Term Cultures with 1000 U/ml of rIL-2[a]

	No. of samples expanded	
Expansion fold	TIL	A-PBL
<2	4	0
2-75	10	22
76-150	4	3
151-500	4	4
>500	9	2
Total pairs studied = 31		

[a]TIL and A-PBL were cultured as described in the legend to Figure 3 and the maximal expansions achieved were compared.

COMPARATIVE EXPANSION OF TIL AND A-PBL

To assess whether or not TIL show better proliferation in rIL-2 than in A-PBL, we grew the tumor- and blood-derived cells of patients with head and neck tumors simultaneously in long-term cultures. As shown in Table 4, TIL from these tumors cultured in the presence of 1000 units/ml of rIL-2 expanded better than A-PBL. The median expansion for 31 TIL preparations from head and neck tumors was 114-fold, in comparison with 31-fold for A-PBL, a significant difference ($p < 0.01$) in cultures maintained for up to 91 days. The best proliferation achieved by TIL from head and neck tumors was 350,000-fold (day 52) and by PBL 1200-fold (day 52). In five instances, A-PBL grew better than TIL in culture. As a general rule, however, TIL achieved higher cell numbers per culture than A-PBL in the presence of the high concentration of rIL-2.

ANTITUMOR EFFECTOR FUNCTION OF EXPANDING TIL

Cultures of TIL were serially monitored for cytotoxicity against natural killer cell (NK)-sensitive and NK-resistant tumor targets during culture. Both cultured and fresh tumor cells were used as targets in 4-hr ^{51}Cr-release assays. As illustrated in Figure 1, TIL from primary solid tumors developed a capability to lyse tumor cell targets with variably good efficiency depending on the target used. The highest levels of antitumor cytotoxicity were achieved relatively late in culture, between days 10 and 50. After this peak,

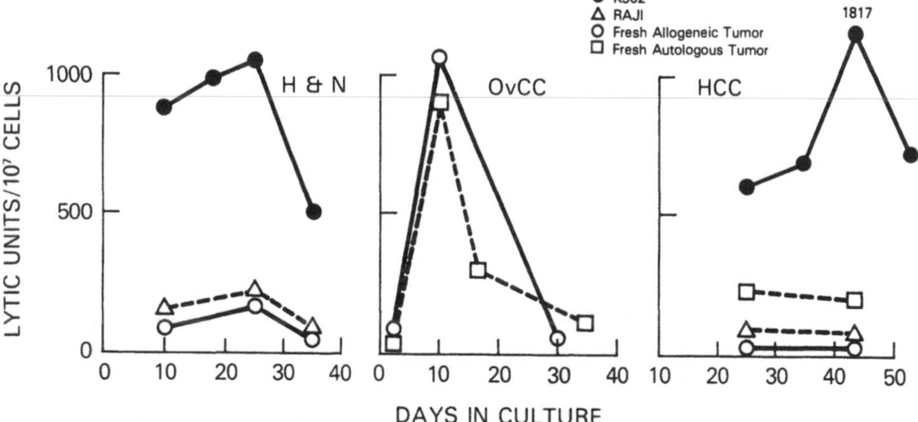

Figure 1 TIL were obtained from three primary solid human tumors: Head and neck squamous carcinoma (H&N), ovarian cell carcinoma (OVCC), and hepatocellular carcinoma (HCC) and expanded in culture in the presence of rIL-2. Cytotoxicity of the expanding cultures was serially monitored against NK-sensitive and NK-resistant cultured and fresh tumor cell targets. Four-hour ^{51}Cr-release cytotoxicity assays were performed under conditions described by us earlier (2). Cytotoxicity was determined as the percentage of specific lysis at four different E/T ratios, and lytic units/10^7 effector cells were calculated using a computer program based on an equation described by Pross et al. (28).

cytotoxicity declined to undetectable levels on days 50 to 80 of growth, when the cultures were generally terminated. The TIL from metastatic tumors showed low levels of cytotoxicity and, in some cases, no cytotoxicity at all during culture with high doses of rIL-2. The reason for poor cytotoxic function of TIL from metastatic disease is not known, but it may reflect the inability of IL-2, even in high concentration, to reverse the inhibitory effects of the tumor environment.

In SCCHN, it was possible to compare antitumor cytotoxicity of TIL with that of A-PBL. In almost all instances, TIL cultures exhibited higher cytolytic function than A-PBL cultures (2). The peak cytotoxicity was achieved considerably later in TIL than A-PBL cultures (2).

These experiments showed that in comparison with LAK cells, TIL from primary solid tumors acquire levels of antitumor cytotoxicity that are higher and reach optimum later in cultures containing 1000 units/ml of rIL-2.

PHENOTYPIC MARKERS ON EFFECTOR CELLS IN TIL CULTURES

The expression of surface markers CD3, CD4, CD8, CD16, and Leu19, which are commonly associated with effector cell function, were serially determined in expanding TIL cultures. Using two-color flow cytometry that facilitates detection of two different markers on the same cell simultaneously, we were able to document shifts in the phenotypic characteristics of TIL outgrowing in the presence of rIL-2 (2,7,12). Although TIL expansion was associated with increases of both $CD3^+$ and $Leu19^+$ cells in culture, it soon became apparent that the $CD3^+$ cells in early TIL cultures were not the same as the $CD3^+$ cells in late cultures (Fig. 2). Furthermore, the peak of antitumor cytotoxicity always corresponded to the peak in frequency of $Leu19^+$ cells in these cultures (2,7,12).

To establish the identity of antitumor effector cells in TIL cultures, we sorted the cultures strained with labeled anti-CD3 and

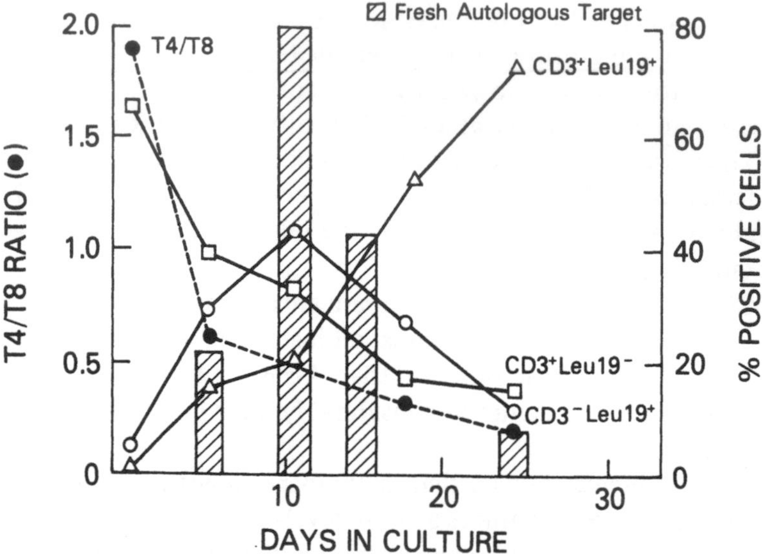

Figure 2 TIL isolated from a primary ovarian carcinoma were established in culture in the presence of 1000 units/ml of rIL-2. The culture was serially monitored for the proportions of CD3⁺ and Leu19⁺ (solid lines) as well as T4⁺ and T8⁺ cells by two-color flow cytometry. The T4/T8 ratios were calculated on the basis of these flow cytometry measurements and are indicated by the broken line. The decreasing T4/T8 ratios were due to the increases in the proportion of T8⁺ cells in the culture. Cytotoxicity was serially measured against fresh autologous tumor cells labeled with ^{51}Cr: Shaded bars indicate cytotoxicity in LU/10^7 cells, as follows (from left to right): 226, 1330, 434, and 84.

anti-Leu19 reagents. The cultures were always sorted at the time of optimal antitumor activity. In Figure 3, the flow cytometry profiles of the mass culture and of sorted populations are presented. The three-way sorts generally yielded populations containing 100% $CD3^+Leu19^-$ cells; 100% $CD3^-Leu19^+$ cells; and 85% to 95% $CD3^+Leu19^+$ cells. The latter population generally contained $CD3^+Leu19^-$ cells as a contaminant. The sorted cell populations were collected, incubated overnight in complete growth medium

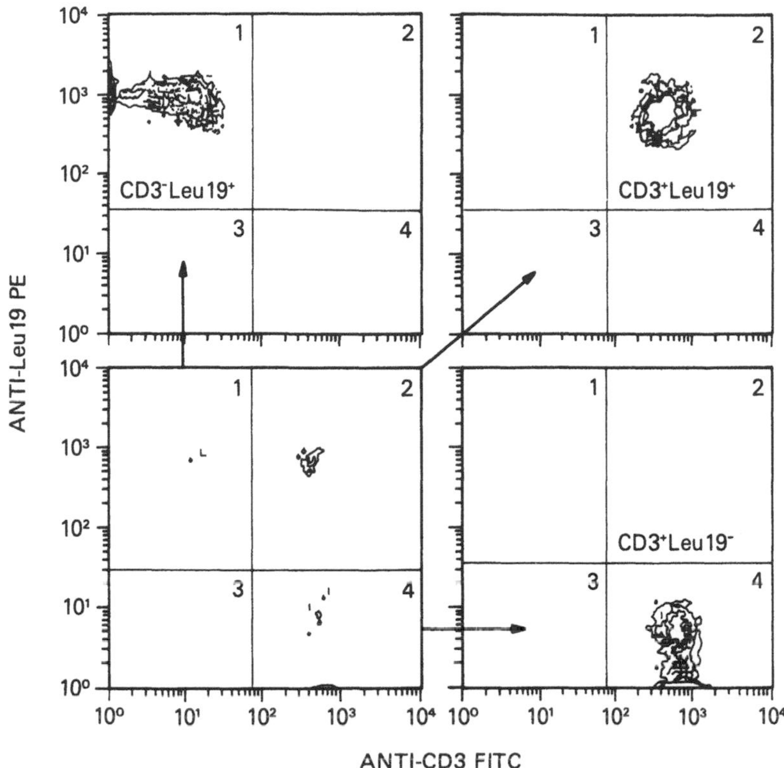

Figure 3 A quantitation and sorting by two-color flow cytometry of a mass culture of TIL from an ovarian carcinoma. The unsorted culture was stained with anti-CD3 FITC and anti-Leu19 PE monoclonal antibodies and was found to contain $CD3^+Leu19^+$ (26%) cells (lower-left quadrant). The cells were first sorted into the $Leu19^+$ and $Leu19^-$populations. Next, the recovered $Leu19^+$ population was further sorted into $CD3^-$and $CD3^+$ fractions. This sorting yielded a $CD3^+Leu19^-$fraction (100% purity), $CD3^+Leu19^+$ population (95% purity), and $CD3^-Leu19^+$ population (100% purity).

Table 5 Effector Cell Function of the TIL Populations Sorted from Mass Cultures Grown in the Presence of High Doses of rIL-2[a]

		Effector Cells			
		$CD3^+Leu19^-$	$CD3^+Leu19^+$	$CD3^-Leu19^+$	
Tumor	Target cell	% Specific lysis at 12:1 E/T ratio			Unsorted
SCCHN (day 16)	K562	44	51	60	56
	Raji	15	30	29	27
	Ovarian CA	9	29	31	13
Ovarian CA (day 10)	K562	36	69	77	76
	Raji	7	48	56	49
	Allo-ovarian CA	3	12	37	19
	Auto-ovarian CA	0	25	36	32

[a]Two representative mass cultures of TIL from a SCCHN and an ovarian carcinoma were checked for antitumor activity in cytotoxicity assays against the target cells listed and sorted after staining with Leu4 and Leu19 monoclonal antibodies as described in the legend to Fig. 3. The sorted effector cell populations were incubated overnight in a complete medium and then assayed for cytotoxicity in 4-hr ^{51}Cr-release assays at the effector/target ratios of 12:1, 6:1, 3:1, and 1.5:1. The percentage spontaneous lysis was <10% for cultured tumor lines and <25% for fresh ovarian carcinoma targets.

and then assayed for antitumor cytotoxicity and, in some cases, for the content of $CD4^+$ and $CD8^+$ cells by immunoperoxidase on cytofuge smears. As shown in Table 5, antitumor cytotoxicity against both autologous and allogeneic tumor cell targets was mediated by $CD3^-Leu19^+$ and $CD3^+Leu19^+$ cells but not by the classic $CD3^+Leu19^-$T lymphocytes. The $Leu19^+$ effector cells were non-MHC-restricted, and antitumor cytotoxocity of sorted $CD3^-Leu19^+$ populations was always greater than that of unseparated cultures tested at the same time (23). The $CD3^+Leu19^+$ effector cells sorted at the time of optimal cytotoxicity in mass cultures expressed the CD8 phenotype.

We interpret the foregoing results to mean that in cultures containing high doses of rIL-2, $Leu19^+$ NK cells and $Leu19^+$ T lymphocytes were the effectors of cell-mediated lysis, whereas the $CD3^+Leu19^-$ T lymphocytes represented the major proliferating cell type in early TIL cultures. The $CD3^+Leu19^+$ cells, which began proliferating extensively between days 10 and 20, soon outgrew the $Leu19^+$ NK cells and $Leu19^-$ T lymphocytes, becoming the major population in late TIL cultures. These cells might exercise suppressive influences in TIL cultures, because their expansion was accompanied by a loss of antitumor cytotoxicity. It is important to note that interactions between different cell populations in mass cultures of TIL determine and modulate the effector cell function, and that purification of effector subpopulations is necessary to generate maximal cytotoxicity in in vitro assays.

INTERLEUKIN-2–EXPANDED TIL ARE NON-MHC-RESTRICTED EFFECTORS

Mass cultures of TIL and sorted effector populations were tested in 4-hr ^{51}Cr-release assays with allogeneic and autologous fresh tumor cell targets. With unseparated cultures, it was possible to assay serially for auto- and allotumor cytotoxicity at several times during expansion of TIL in the presence of rIL-2. These experiments required the availability of fresh or cryopreserved autologous tumor cells and were performed in our laboratory with

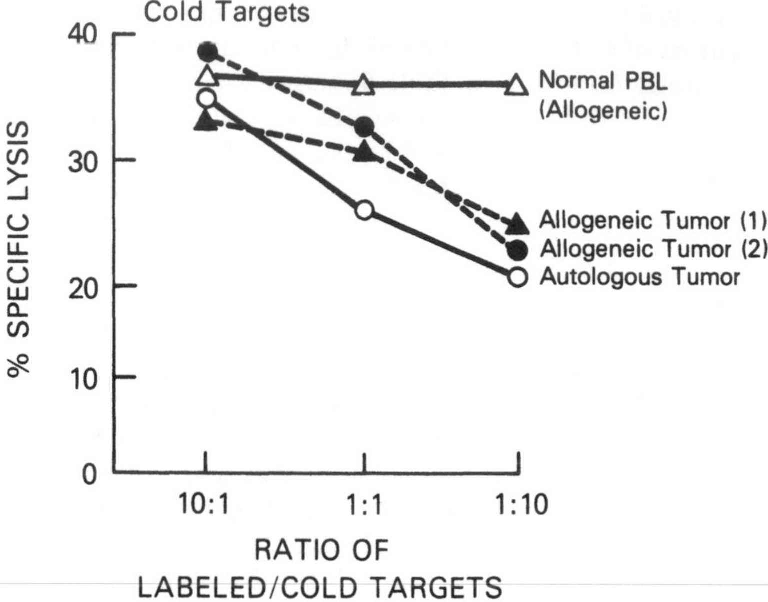

Figure 4 The cold-target inhibition experiment was performed with autologous and allogeneic ovarian carcinoma cells as targets and IL-2-grown TIL from the ovarian tumor as effectors. Graded doses of cold targets (normal PBL as controls, allogeneic ovarian tumors 1 and 2, autologous ovarian tumor) were added to the effectors at 0.1, 1, and 10 times the labeled target cell number. A fixed number of labeled autologous target cells (5×10^3) were then added to each well. The targets and effectors were incubated for 30 min at 37°C. The cytotoxicity of the effector cells against the labeled autologous tumor cells (at E/T ratio of 50:1) was determined. Cold allogeneic tumor targets from two different ovarian tumors competed nearly as well as autologous targets for the effector cells.

TIL from primary ovarian carcinomas. Specificity of TIL for autologous tumor cells was never demonstrated during culture (12). To confirm the lack of specificity, the cold-target inhibition was performed as indicated in Figure 4. Unlabeled allogeneic carcinoma competed as effectively as labeled autologous ovarian carcinoma targets for the effectors generated in TIL cultures (see Fig. 4).

In our hands, the CD3⁻Leu19⁺ effector cells mediated non-MHC-restricted cytolytic activity, and although preferential lysis for autologous tumor cell targets was seen at a single point in some cultures, it was transient and accompanied by the ability to lyse allogeneic tumor targets of the same or different histologic type.

CONCLUSIONS

Clinical trials employing IL-2-expanded TIL in adoptive transfers to patients with metastatic cancers refractory to conventional therapy have begun in at least two centers in the United States. The preliminary results from both of these trials have been published and are not especially encouraging (4,24). Recently, Rosenberg's group described a procedure for large-scale expansion of human TIL in 1000 units/ml of rIL-2 that occasionally generated more than 10^{10} lymphocytes (24). These expanded TIL cultures were said to be almost pure cultures of activated T lymphocytes, the phenotype of which was $CD3^+CD8^+HLA\text{-}DR^+$ (25). Although these investigators acknowledged that lytic characteristics of cultured TIL change over time, the outgrowth of $CD3^+$ T lymphocytes was equated with optimal antitumor activity (25).

Our interpretation of events occurring in long-term cultures of IL-2-expanded TIL is different. By purifying the antitumor effectors from mass cultures of IL-2-expanded TIL, we were able to show that relatively minor populations of $CD3^-Leu19^+$ cells (< 10%) and $CD3^+Leu19^+$ cells (< 10%) were responsible for all antitumor cytotoxicity, whereas the predominating $CD3^+Leu19^-$ T cells (80%) had little if any cytotoxicity. Furthermore, a decline with time of cytotoxic activity in TIL cultures was associated with a disappearance of the two effector populations and expansion of the $CD3^+Leu19^+$ population that was unable to mediate antitumor cytotoxicity (26).

Several conclusions can be drawn from our observations: (a) TIL from human solid tumors growing in the presence of 1000 units/ml of rIL-2 have some advantage in terms of proliferation and antitumor cytotoxicity over LAK cultures; (b) this advantage is counterbalanced by a relative difficulty of obtaining TIL and the fact that TIL from metastatic tumors may fail to grow or have little or no cytotoxicity when expanded; (c) two minor populations of NK cells and Leu19$^+$ T cells mediate antitumor cytotoxicity in TIL cultures grown in high doses of rIL-2, and the same two populations are present and active in cultures of A-PBL; (d) neither TIL nor A-PBL–generated effectors are specific for autologous tumor in the presence of 1000 units/ml of rIL-2.

The wisdom of clinical trials with TIL expanded in high doses of rIL-2 needs to be reevaluated at this time. The key question concerns the presence of the MHC-restricted, auto–tumor-specific effectors in TIL preparations. In view of experiments in murine models of metastatic disease (1,20) and a very recent in vitro demonstration of CTL in human carcinomas of the bladder and larynx (27), the likelihood of obtaining CTL-specific for autologous tumor is good. However, it may be necessary to search for culture conditions more favorable to tumor-specific T lymphocytes, if their presence in TIL is to be confirmed and their in vitro expansion achieved. Because high doses of rIL-2 appear to induce non–MHC-restricted effector cells, it may be necessary to consider altering IL-2 concentrations, adding other cytokines or activators, or removing suppressor influences that are possibly mediated by suppressor cells or by tumor-derived products. Thus, considerable in vitro work must be invested to generate TIL that are qualitatively different from other effectors of non–MHC-restricted cytotoxicity. The availability of such qualitatively different CTL should provide the strongest arguments for the preferential use of TIL in adoptive immunotherapy of cancer.

REFERENCES

1. Rosenberg S A, Spiess P, Lafreniere R. A new approach to the adoptive immunotherapy of cancer with tumor infiltrating lymphocytes. Science 1986;233:1318–1321.
2. Heo D S, Whiteside T L, Johnson J T, Chen K, Barnes E L, Herberman R B. Long-term interleukin-2–dependent growth and cytotoxic activity of tumor infiltrating lymphocytes from

human squamous cell carcinomas of the head and neck. Cancer Res 1987;47:1-10.
3. Muul L M, Spiess P J, Director E P, Rosenberg S A. Identification of specific cytolytic immune responses against autologous tumor in humans bearing malignant melanoma. J Immunol 1987;138:989-995.
4. Topalian S L, Solomon D, Avis F P, et al. Immunotherapy of patients with advanced cancer using tumor-infiltrating lymphocytes and recombinant interleukin-2: A pilot study. J Clin Oncol 1988;6:839-853.
5. Whiteside T L, Heo D S, Takagi S, Johnson J T, Iwatsuki S, Herberman R B. Cytolytic antitumor effector cells in long-term cultures of human tumor-infiltrating lymphocytes in recombinant IL-2. Cancer Immunol Immunother 1987;26: 1-10.
6. Ioachim H L. The stroma reaction of tumors: An expression of immune surveillance. J Natl Cancer Inst 1979; 57:465-475.
7. Whiteside T L, Miescher S, Hurlimann J, Moretta L, von Vliedner V. Separation, phenotyping and limiting dilution analysis of lymphocytes infiltrating human solid tumors. Int J Cancer 1986;37:803-811.
8. Wolf G T, Hudson J L, Peterson K A, Miller H L, McClatchey K D. Lymphocyte subpopulations infiltrating squamous carcinomas of the head and neck: Correlations with extent of tumor and prognosis. Head Neck Surg 1986;95:142-152.
9. Svennevig J L, Lunde O C, Holter J, Bjorgsvik D. Lymphoid infiltration and prognosis in colorectal carcinoma. Br J Cancer Res 1984;49:375-377.
10. Kreider J W, Bartlett G L, Butkiewicz B L. Relationship of tumor leukocytic infiltration to host defense mechanisms and prognosis. Cancer Metastasis Rev 1984;3:53-74.
11. Brocker E B, Kolde G, Steinhausen D, Peters A, Macher E. The pattern of the mononuclear infiltrate as a prognostic parameter in flat superficial spreading melanomas. J Cancer Res Clin Oncol 1984;107:48-52.
12. Heo D S, Whiteside T L, Kembour A, Herberman R B: Role of Leu19 (NKH1)-positive effector cells in mediating autol-

ogous and allogeneic tumor cell lysis in rIL-2-activated cultures of lymphocytes infiltrating human ovarian tumors. J Immunol 1988; 140:4042–4049.

13. Itoh K, Tilden A B, Balch C N. Interleukin-2 activation of cytotoxic T lymphocytes infiltrating into human metastatic melanomas. Cancer Res 1986; 46:3011–3017.
14. Allavena P, Zanaboni F, Rossini S, Merendino A, et al. Lymphokine-activated killer activity of tumor-associated and peripheral blood lymphocytes isolated from patients with ascites ovarian tumors. J Natl Cancer Inst 1986; 77:863–868.
15. Whiteside T L, Miescher S, MacDonald R H, von Fliedner V. Separation of tumor infiltrating lymphocytes from human solid tumors. A comparison of velocity sedimentation and discontinuous density gradients. J Immunol Methods 1986; 90:221–233.
16. Vose B M. Separation of tumor and host cell populations from human neoplasms. In Reid E, Look G M W, Moore D J (eds): Cancer Cell Organelles. Chichester, Ellis Harwood Ltd, 1982.
17. Miescher S, Whiteside T L, Carrel S, von Fliedner V. Functional properties of tumor infiltrating and blood lymphocytes in patients with solid tumors: Effect of tumor cells and their supernatants on proliferating responses of lymphocytes. J Immunol 1986; 136:1899–1907.
18. Herberman R B, Holden H T, Varesio L, Taniyama T, et al. Immunologic reactivity of lymphoid cells in tumors. In Hanna N G, Witz I P (eds): Contemporary Topics in Immunobiology. New York, Plenum Press, 1980:61–76.
19. Brunner K T, MacDonald H R, Cerottini J-C. Quantitation and clonal isolation of cytotoxic T lymphocyte precursors selectively infiltrating murine sarcoma virus-induced tumors. J Exp Med 1981; 154:362–373.
20. Shu S, Chou T, Rosenberg S A. In vitro sensitization and expansion with viable tumor cells in interleukin-2 in the generation of specific therapeutic effector cells. J Immunol 1986; 136:3891–3898.

21. Roth J A, Grimm E A, Gupta R K, Ames R S. Immunoregulatory factors derived from human tumors: Immunologic and biochemical characterization of factors that suppress lymphocyte proliferative and cytotoxic responses in vitro. J Immunol 1982; 128:1955–1962.
22. Moy P M, Holmes E C, Golub S H. Depression of natural killer cytotoxic activity in lymphocytes infiltrating human pulmonary tumors. Cancer Res 1985; 45:57–60.
23. Whiteside T L, Heo D S, Takagi S, Herberman R B. Characterization of novel antitumor effector cells in long-term cultures of human tumor-infiltrating lymphocytes. Transplantation Proc 1988; 20:347–350.
24. Kradin R L, Boyle L A, Preffer F L, Calahan R J, et al. Tumor-derived interleukin-2-dependent lymphocytes in adoptive immunotherapy of lung cancer. Cancer Immunol Immunother 1987; 24:76–85.
25. Topalian S L, Muul L M, Solomon D, Rosenberg S A. Expansion of human tumor-infiltrating lymphocytes for use in immunotherapy trials. J Immunol Methods 1987; 102:127–141.
26. Takagi S, Chen K, Schwarz R, Iwatsuki S, Herberman R B, Whiteside T L. Functional and phenotypic analysis of tumor-infiltrating lymphocytes isolated from human liver tumors and cultured in recombinant IL-2. Cancer 1988; (in press).
27. Cozzolino F, Torcia M, Carossino A M, Giordani R, et al. Characterization of cells from invaded lymph nodes in patients with solid tumors. Lymphokine requirement for tumor-specific lymphoproliferative response. J Exp Med 1987; 166:303–318.
28. Pross H F, Baines N T, Rubin P, Shragge P, Patterson N S. Spontaneous human lymphocyte mediated cytotoxicity against tumor target cells. IX. The quantitation of natural killer cell activity. J Clin Immunol 1981; 1:51–63.

10

Tumor-Derived Activated Cells: Culture Conditions and Characterization

JAMES R. MALECKAR, COLLEEN S. FRIDDELL, WALTER M. LEWKO, and JOHN R. YANNELLI
Biotherapeutics, Inc., Franklin, Tennessee

WILLIAM H. WEST
Biotherapeutics, Inc., Memphis, Tennessee

ROBERT K. OLDHAM
Biological Therapy Institute, Franklin, Tennessee

For over 30 years it has been known that tumor-bearing hosts can be immunized against their own tumor (1). This suggest that tumor-specific cytotoxic T lymphocytes (CTL) and T-helper cells are generated in cancer patients. Lymphocyte infiltration of tumor tissue was noticed as early as 1907 (2). Later, Miwa recognized that those cancer patients who had many tumor-infiltrating lymphocytes had a better prognoses than those patients with a smaller degree of infiltration (3). These findings suggested that the lymphocytes found within tumors may play a major role in the host defense against the neoplasia. The assumption made is that in vivo the tumor has been enriched for lymphoid cells having antitumor reactivity. On the other hand, it is apparent that these cells are "blocked" or not available in sufficient numbers, because the tumor continues to grow, in spite of their presence.

The in vivo relevance of the antitumor activity of T cells was first shown in murine models (4–6). Early work by Cheever et al.

(reviewed in Ref. 5) elucidated the specific antitumor activity of peripheral and splenic T cells in a murine model. Treves et al. (4) demonstrated that the injection of in vitro tumor-sensitized lymphocytes into mice bearing an otherwise lethal Lewis lung carcinoma significantly increased survival rate. Rosenberg and coworkers (6) observed that lymphocytes that had been expanded from tumors had a much greater potential in reducing tumor loads in mice than LAK cells.

Recently, a number of groups have focused their attention on the practical aspects of expanding and activating the lymphocytes found in tumors for use in cancer biotherapy (7–13). In addition to our studies (13), at least two other laboratories have already initiated pilot studies for the treatment of cancer patients with tumor-derived lymphocytes (11,12).

In this chapter, results on the expansion and characterization of lymphocyte populations from human tumors are presented. Finely minced tumor tissue from cancer patients was cultured in the presence of recombinant interleukin-2 (rIL-2). The lymphocytes that grew out of the tumor cell cultures were primarily T cells and were termed *tumor-derived activated cells* (TDAC). Although the phenotypic or subpopulation composition of TDAC varied, specific lytic activity was frequently observed and was always correlated with TDAC populations enriched for $CD8^+$ lymphocytes. Continued expansion and maintenance of specificity up to the cell levels needed for therapeutic treatment was accomplished through periodic restimulation of TDAC with autologous tumor cells. Clinical trials have been initiated with TDAC as a source of biotherapy, and preliminary results will be discussed.

INITIATION OF THE TDAC CULTURES

The TDAC cultures were initiated from tumor chunks and cells (lymphoid and tumor) released in the mincing process. We feel that the advantage of this system is that the in vivo microenvironment is preserved in the tissue chunks. At the same time, this method allows a proper interaction of lymphocytes, tumor cells, and accessory cells. Precedent has been shown for the use of small tissue fragments in the successful culture of specific lymphoid cells (14,15). In addition, the amount of manipulation in preparing these cultures is greatly reduced compared with those methods that employ extensive enzymatic digestion to obtain a single-cell suspension. Enzymes are also known to alter the expression of certain tumor antigens and may affect receptor function of lymphocytes. The tumor preparations (chunks and cells) were initiated in cell culture in a medium that contained 10% heat-inactivated human serum and 1000 units/ml r-IL2. The T75-cm^2 flasks were seeded with 0.5 to 1.0 g of tumor preparation. Cultures were incubated 5 to 7 days at 37°C in 5% CO_2 in air environment. The cultures were then washed and reestablished at a lymphoid cell density of 5×10^5 cells/ml. At this point, cultures were supplemented with 20% lymphokine-activated killer cell (LAK)-conditioned medium (see Chap. 12). We observed that

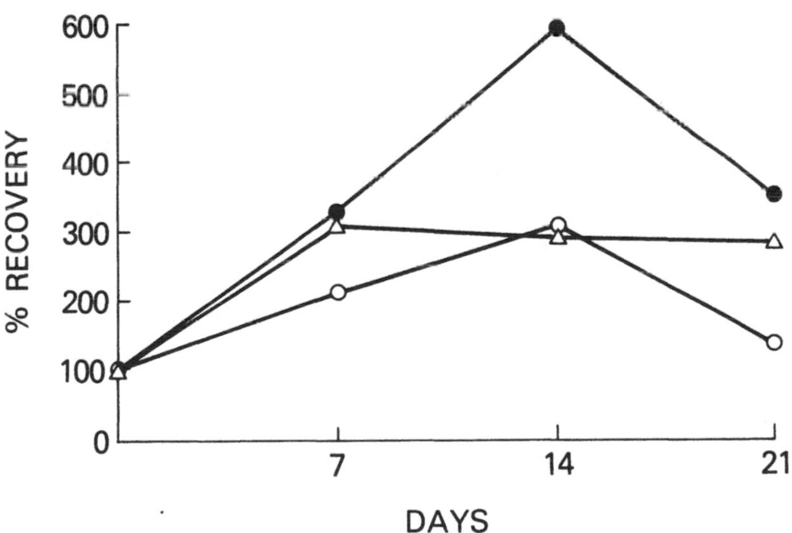

Figure 1 The effect of LAK-conditioned medium on the expansion of TDAC. The growth rates of TDAC initiated in the presence, closed circles, or absence, open circles, of 20% (v/v) LAK-conditioned medium as well as supplemented 1 week after culture initiation, triangles, were observed for the first 3 weeks of culture.

the addition of LAK-conditioned medium was a requirement for long-term culture of TDAC (Fig. 1). In addition, optimal enhancement was observed when the LAK-conditioned medium was first added 1 week after the initiation of cell cultures.

TUMOR ACQUISITION, PROCESSING, AND PRESERVATION

Tumor specimens enter the TDAC program through our Tumor acquisition, processing, and preservation (TAPP) program. The TAPP program is designed to acquire tumor specimens from a cancer patient's own oncologist and surgical team. Tumors are transported in sterile transport medium in an insulated container with cold packs (cool, not frozen). When handled in this manner, tumor specimens maintain viability and outgrowth potential for at least 24 hr.

In the present study, the tumors were mechanically dissected into finely minced chunks ranging from 1 to 3 mm^3. Tumors prepared in this manner consisted of tissue chunks, as well as a single-cell suspension of tumor cells and lymphoid cells released in the mincing process. Tumor cell cultures were initiated using the released tumor cells or enzymatically derived cells obtained from the tumor chunks. In brief, enzyme digests of the chunks were accomplished with Worthington type-II collengenase (2 mg/ml) in RPMI-1640 medium, supplemented with 5% FCS for 1 hr at 37°C (Lewko et al., manuscript in preparation). Development of tumor cell lines required, in general, 3 to 5 months. However, usable cells were sometimes obtained from certain cultures within 8 weeks. The tumor cell lines were used as either an antigen source for TDAC restimulation or as target cells for specificity analysis.

After the initial cell harvesting and feeding, TDAC cultures were passaged every 5 to 7 days and reestablished as described earlier. Determination of the day of cell feeding depended upon the TDAC reaching a lymphocyte density between 1.5 and 2.0×10^6 cells/ml. Interestingly, through this harvesting and feeding process, chunks and cellular debris disappeared after 3 weeks. It is at this point, we found it necessary to restimulate the TDAC.

ANTIGEN RESTIMULATION OF TDAC CULTURES

Crucial to the culturing process was the continued effort to provide the TDAC a source of antigen as diverse as the original biopsy. It is well established that tumors are very heterogeneous; therefore, we feel that the immune response we generate in vitro should match this heterogeneity. Proliferation and generation of T-cell specificity was accomplished through biweekly addition to

Table 1 The Effect of Antigen Stimulation on the Growth (Panel A) and Lytic Activity (Panel B) of TDAC

Panel A

Population	% Recovery -Antigen	% Recovery +Antigen[a]	Stimulation index[b]
1	105	457	4.4
2	131	160	1.2
3	380	526	1.4
4	600	788	1.3
5	93	378	4.1

Panel B

	Lytic units/10^7 cells - Antigen	Lytic units/10^7 cells +Antigen[a]	Ratio[c]
A	459	1188	2.6
B	1033	1024	1.0
C	100	799	8.0

[a]Tumor cells were irradiated and included in the TDAC culture at a tumor cell to TDAC ratio of 1:50.

[b]Stimulation index = $\frac{+\text{ Ag (\% recovery)}}{-\text{ Ag (\% recovery)}}$

[c]Ratio = the ratio of + (LU) to - Ag (LU)

the TDAC cultures of irradiated cryopreserved tumor cells from the original biopsy or irradiated cells from an autologous tumor cell line (Table 1). We prefer using the original tumor in order to stimulate the TDAC with the same antigenic determinants as those that were present during the initiation of the TDAC cultures. In addition, these chunks also contained antigen-presenting cells and other lymphokine-producing cells important in the process. However, autologous tumor cells from a cultured line usually were capable of stimulating the TDAC. The use of cultured cells is primarily useful when the size of the biopsy specimen is small. We are currently investigating other means of stimulating the TDAC by using xenografts as well as other methods. Cultures that had not had antigen added were found to lose both cytolytic and proliferative activity, and died after 3 to 4 weeks.

LARGE-SCALE CULTURING, EXPANSION, AND HARVESTING OF TDAC

When TDAC reach a level of 1×10^9 cells or higher, cultures were established in Fenwall PL732 bags. The techniques associated with

Table 2 Comparison of TDAC Cultured in the Presence of Either LAK-Conditioned Medium or TDAC-Conditioned Medium

		Source of conditioned medium	
Experiment[a]	Passage	LAK	TDAC
1	1	304[b]	336
	2	776	752
	3	795	680
2	1	433	567

[a]TDAC from an established culture were supplemented with 25% LAK conditioned medium or TDAC-conditioned medium (from the previous week's culture).

[b]Results are expressed as percent recovery.

the use of bags and the automated procedure for LAK cultures applicable to TDAC have been described in detail by Yannelli et al. (17) and are reviewed in Chapter 12. At this point, the TDAC cultures compromised predominantly T cells, and the growth medium from the previous week's culture was shown to enhance the TDAC growth as effectively as LAK-conditioned medium (Table 2). Thus, reliance on LAK-conditioned medium decreased. Subsequent passages consisted of diluting the contents of each bag into bags containing new medium at a volume necessary to reduce cell density back to 5×10^5 cells/ml, as determined by sample precounts. In general, this old bag/new bag ratio ranged from 1:3 to 1:6. The final harvest of TDAC for reinfusion was done when the numbers of TDAC reach 5×10^{10} cells or higher. We are currently developing other methodologies for TDAC expansion that reduce the amount of technical time involved in the process. In particular, bioreactors are attractive.

CHARACTERIZATION OF TDAC CELLS

A panel of nine TDAC populations derived from melanoma patients was used to analyze cell surface phenotype and lytic specificity 3 to 4 weeks after the initiation of TDAC cultures. Almost all (>90%) of the cells expanded from the tumors were T lymphocytes (Table 3). The NK cells and LAK cells, defined by the expression of the NKH-1 marker and the coexpression of NKH-1 and Tal markers (16), respectively, made up less than 2% of the lymphocyte populations. Four of nine TDAC populations examined consisted primarily of $CD8^+$ lymphocytes (TDAC B, C, D, and E). One of the cultures (TDAC A) contained a substantial mixture of $CD8^+$ and $CD4^+$ lymphocytes. The remaining four cultures (TDAC F, G, H, and I) consisted primarily of $CD4^+$ lymphocytes. Interestingly, in one patient, lymphocytes expanded from tumors excised from two different sites (TDAC E and F) were markedly different in phenotype.

The cytolytic activity of TDAC populations from several patients was assessed against a panel of tumor targets (see Table 3).

Table 3 Cytolytic Activity of TDAC Generated from 9 Melanoma Patients

			Tumor target[a]		
Population	T4/T8[b]	K562	Daudi	Autologous	Allogeneic
TDAC A	40/60	26	0	1188	27
LAK[c]	ND[d]	4284	3643	332	55
TDAC B	12/80	1449	284	1033	10
LAK	ND	4550	3957	458	114
TDAC C	3/96	12	13	1363	3
LAK	ND	955	443	55	ND
TDAC D	1/97	57	0	799	136
LAK	ND	8520	1032	57	678
TDAC E	3/97	16	2	372	0
LAK	ND	1077	552	92	ND
TDAC F	98/2	23	2	2	0
LAK	ND	1077	552	92	ND
TDAC G	84/14	6	0	0	0
LAK	ND	4549	3956	114	458
TDAC H	78/11	15	1	38	0
LAK	ND	2269	1636	832	45
TDAC I	98/1	71	16	0	ND
LAK	ND	3227	2724	758	ND

[a]Results expressed as lytic units per 10^7 cells.
[b]% of CD4+ (T4) and CD8+ (T8) cells in each culture.
[c]Allogeneic LAK cells.
[d]ND = not done.

Usually, allogeneic LAK cells were used as a control to determine the target cell susceptibility to lysis. The TDAC populations A, C, and E killed only autologous tumor targets. The TDAC B cells had strong lytic activity against the autologous tumor; however, some activity was evident against the NK-sensitive target K562 and the LAK-sensitive target Daudi. The TDAC D population exhibited lytic activity against both the allogeneic and autologous tumor target, although the lysis against autologous tumor was considerably greater. The lymphocytes expanded from TDAC F, G, H, and I (primarily T4 cells) did not have any significant lytic activity against any of the targets tested. When autologous tumor lysis by TDAC was observed, the level of cytolytic activity was always at least three times higher than that observed for LAK-mediated lysis. This difference may be due to a higher concentration of killer cells within the TDAC cultures, as opposed to LAK cultures, or it may reflect higher lytic activity by TDAC on a per cell basis. It should be noted that TDAC E and F differed both in their lytic activity and their phenotype, although they originated from the same patient. Therefore, cultivation of TDAC from one biopsy specimen from a patient does not necessarily indicate that cultures initiated from other metastatic sites would be successful or would have the same biological activities.

GROWTH OF TDAC FOR CLINICAL TRIALS

In the development of our clinical TDAC program, we have examined tissues from a variety of different tumor types. Our criteria for tumor selection did not include restrictions on specimen size (range: 0.3 g-> 100 g) or viability of leukocyte infiltrate (range: 0%–80%). Our overall success rate in growing at least 1×10^{10} lymphocytes from a tumor biopsy is now 80% (Table 4), with a range of 60% for adenocarcinoma to 100% for breast tumors. We feel that this success rate should reflect that which would be observed in the general cancer population. Along these lines we have developed a program termed *assessment for biotherapy of cancer* (ABC) that yields information on the feasibility of growing TDAC for any patient, even if the patient is not ready for treatment. The TDAC are expanded to a certain level and cryopreserved for later use. They can then be thawed and expanded to a therapeutic level when required. We have been routinely successful in recovering TDAC from cryopreserved specimens, and this method has now been utilized for the preparation of cells for two patients.

Table 4 Expansion of TDAC from Various Tumor Types

Tumor type	No. of patients	% Positive growth
Melanoma	28	82
Ovarian	7	86
Renal cell	4	75
Adenocarcinoma	5	60
Colon	10	90
Breast	4	100
Other	16	63
Total	74	80

We have currently prepared TDAC for nine cancer patients (Table 5). These patients were represented by four different tumor types. There was no correlation between the tumor types or number of lymphocytes grown. One patient received predominantly T4 lymphocytes, six patients received predominantly T8, and two patients received a mixture of the two. In three of the nine patients, a population of cells was derived that was specifically cytolytic toward autologous tumor cells. In three other patients, lymphocytes were shown to kill tumor targets in a redirected lysis system. In brief, this system uses antibodies called heteroconjugates, which consist of monoclonal antibodies against the CD3 determinant covalently linked to monoclonal antibodies against a tumor-associated antigen. These heteroconjugates bridge the killer cell with the unrelated target. Thus, the cytolytic potential of the killer population can be measured in the absence of the specific tumor target. In the patients that received predominantly T4 cells, specificity was suggested by increased proliferation of the TDAC upon addition of autologous tumor cells.

Of the nine patients treated, two of the first three patients, both with metastatic melanoma, have shown biological activity with tumor reduction. Subsequent patients are still undergoing

Table 5 Summary of Laboratory Data from the First 9 Patients Treated with TDAC

Patient	Tumor type	No. of cells infused ($\times 10^{10}$)	Phenotype T4/T8[a]	Cytolytic activity[b]
1	Melanoma	6.0	50/50	+
2	Melanoma	5.5	20/80	+
3	Melanoma	23.0	2/98	-
4	Melanoma	27.0	2/98	-
5	Lung	13.0	1/99	+
6	Colon	19.0	7/93	+
7	Colon	7.6	92/8	-
8	Colon	10.0	27/73	+
9	Ovarian	13.0	60/40	+

[a]Results expressed as percent positive.
[b]Cytolytic activity was measured either directly (i.e., autologous tumor target) or indirectly (i.e. the heteroconjugate system).

treatment or being observed for clinical effects. It is anticipated that 20 to 25 patients will need to be treated, with a follow-up of at least 3 months posttreatment before sufficient clinical data will be available for publication.

We have been very interested in using combination therapies of TDAC and LAK protocols. A major advantage includes the generation of both specific (TDAC) and nonspecific (LAK) cytolytic effector cells. When reevaluating the data presented in Table 4, it was observed that growth of TDAC from patients treated with LAK–rIL-2, before removal of the tumor biopsy, occurred 93% of the time. This is a substantial increase above the growth rate observed from patients having no prior LAK–rIL-2 treatment (78%). In addition, phenotypic analysis suggests that the nodules following LAK–rIL-2 treatment contained a larger percentage of activated T cells (data not shown).

CONCLUSIONS

Given the accumulated data from nearly 20 years of immunotherapy, it is now clear that T lymphocytes can mediate significant antitumor responses. A limiting factor has always been the ability to expand T cells, whether they were derived from the peripheral blood, spleen, or tumor. The recent availability of rIL-2 has demonstrated the feasibility of expanding T cells from any of these sources and has demonstrated clinical efficacy of these cells as antitumor effectors in murine models.

Concomitant with the demonstration that rIL-2 can provide the growth stimulus for T cells, which are then useful in treating cancer, the LAK cell technology came to the clinic demonstrating that peripheral blood cells with a function distinct from T cells could be cultured, expanded, and reinfused in patients with significant clinical effects. Although the most promising results have been in patients with melanoma and renal cancer, antitumor effects have been seen in patients with a wide variety of cancers, even those with bulky tumors. This form of adoptive cellular biotherapy has confirmed that an expanded and activated cell population from the cancer research laboratory can provide a method by which clinicians can effectively treat advanced cancer.

In tumor biopsy specimens, the infiltrating lymphocytes have been recognized and are known to be cytolytically active for many years. As in the mouse, the major limiting factors were the ability to culture large numbers of these infiltrating cells and the limited understanding of the tumor antigens involved for T-cell stimulation. The TDAC technology, as described in this chapter, makes it clear that the technical problems of T-cell expansion are now being solved. The restimulation by antigen appears to be providing the ongoing antigen stimulation needed to maintain selective killing of tumor cells. Various factors in the medium that support and enhance growth and T-cell activation are being defined. Thus, the components are now available to develop a broad attack on advanced cancer using this laboratory-based technology of tumor-derived activated cell stimulation, expansion, and therapy.

Although these results and the results from a few other laboratories are preliminary, they point to the dramatic change in technology that has allowed the cancer research laboratory to be a substantial component in evolving new clinical approaches to cancer treatment. As is apparent from this chapter and previous chapters on LAK cell activation, the laboratory scientists and their technical expertise have become a major component in the design and conduct of clinical trials using adoptive biotherapy. As these studies continue, it will be essential to conduct experiments to determine which T-cell population is therapeutically most effective. The role of factors in the medium for expansion and activation will be critically important to understand. The role of antigen stimulation is also basic to further progress with this technology. Tumor cell chunks, tumor cell cultures, nude mouse xenografts, or purified antigen, all represent potential sources of repeated antigenic stimulation. All of these techniques are laboratory based, and it is only with close and effective communication between the laboratory scientists and the clinician that rapid and effective translation of these technologies to the patient will occur.

REFERENCES

1. Prehn RT, Main JM. Immunity to methycholanthrene-induced sarcomas. J Natl Cancer Inst 1957; 18:769.
2. Handley WS. The pathology of melanotic growths in relation to their operative treatment. Lancet 1907; 1:927.
3. Miwa H. Identification and prognostic implications of tumor infiltrating lymphocytes. Acta Med O Kayama 1984; 38:215.
4. Treves AJ, Cohen IR, Feldman M. Brief Communication: immunotherapy of lethal metastases by lymphocytes sensitized against tumor cells in vitro. J Natl Cancer Inst 1975; 54:777.
5. Cheever MA, Greenberg PD, Fefer A. Potential for specific cancer therapy with immune T lymphocytes. J Biol Response Mod 1984; 3:113.
6. Rosenberg SA, Spiess P, Lafreniere R. New approach to the

adoptive immunotherapy of cancer with tumor-infiltrating lymphocytes. Science 1986; 233:1318.
7. Yron I, Wood TA Jr, Spiess PJ, Rosenberg SA. In vitro growth of murine T cells. V. The isolation and growth of lymphoid cells infiltrating syngeneic solid tumors. J Immunol 1980; 125:238.
8. Topalian SL, Muul LM, Solomon D, Rosenberg SA. Expansion of human tumor infiltrating lymphocytes for use in immunotherapy trial. J Immunol 1987; 102:127.
9. Muul LM, Spiess PJ, Director EP, Rosenberg SA. Identification of specific cytolytic immune responses against autologous tumor in humans bearing malignant melanoma. J Immunol 1987; 138:989.
10. Itoh K, Tilden AB, Balch CM. Interleukin 2 activation of cytotoxic T-lymphocytes infiltrating into human metastatic melanomas. J Immunol 1986; Cancer Res 1986; 46:3011.
11. Kradin RL, Boyle LA, Preffer FI, Callahan RJ, Barlaikovach M, Strauss HW, Dubinett S, Kurnick JT. Tumor derived interleukin-2-dependent lymphocytes in adoptive immunotherapy of lung cancer. Cancer Immunol Immunother 1987; 24:76.
12. Belldegrun A, Muul LM, Rosenberg SA. Interleukin-2 expanded tumor-infiltrating lymphocytes in human renal cell cancer: Isolation, characterization, and anti-tumor activity. Cancer Res 1988; 48:206.
13. Maleckar JR, Friddell CS, Price WJ, Thurman GB, Lewko WP, West WH, Oldham RK, Yannelli JR. Activation and expansion of tumor derived activated cells (TDAC). 1988 (Submitted for publication).
14. Klinman NR. The mechanism of antigenic stimulation of primary and secondary precursor cells. J Exp Med 1972; 136: 241.
15. Klinman NR, Wylie DE, Cancro MP. Mechanisms that govern repertoire expression. In Fougereau M, Dausset J (eds): Immunology 1980 (Proc 4th International Congress of Immunology). London, Academic Press, 1980:123.
16. Yannelli JR, Desch C, Shults K, Houston J, Stelzer G. Characterization of human lymphokine activated killer cells

(LAK): Cytotoxicity and cell surface phenotype. Fed Proc 1987;46:483.

17. Yannelli JR, Thurman GB, Dickerson SG, Mrowca A, Sharp E, Oldham RK. An improved method for the generation of human lymphokine activated killer cells. J Immunol Methods 1987;100:137.

11

Clinical/Technical Challenges in Adoptive Cellular Immunotherapy (ACI): The Role of Cytapheresis

LEOCADIO V. LACERNA
National Cancer Institute, Frederick, Maryland

JEANE HESTER
M.D. Anderson Hospital, The University of Texas System Cancer Center, Houston, Texas

ALVARO A. PINEDA and EDWIN BURGSTALER
Mayo Clinic Blood Bank, Mayo Clinic, Rochester, Minnesota

HARVEY G. KLEIN
Warren G. Magnuson Clinical Center, National Institutes of Health, Bethesda, Maryland

A recent approach to the treatment of human cancer has been termed adoptive cellular immunotherapy (ACI) involves transfer of autologous, in vitro-activated immune reactive cells such as lymphocytes or monocytes are passively transferred to tumor-bearing hosts in order to mediate antitumor responses by direct or indirect mechanisms. A substantial research effort is currently underway which is designed to treat patients with incurable cancer with these novel regimens. Such approaches, to date, include clinical trials utilizing lymphokine-activated killer (LAK) cells, tumor-infiltrating lymphocytes (TIL), and gamma interferon-(IFN-γ) activated killer monocytes (AKM). More than 30 centers are now involved in such studies in the United States alone. Clinical progress in this area has been well summarized in Chapters 3-6, 8, and 10.

This chapter focuses on the role of cytapheresis in adoptive cellular immunotherapy. In ACI specific mononuclear leukocytes are collected from the cancer patient and their antitumor activity is "up-regulated" in vitro prior to reinfusion. Several general principles are applicable to the harvesting and purification of these effector cells: (1) When possible, isolate a single purified cytotoxic effector cell population for initial clinical trials in the ACI setting. Mixtures of different effector cell types may have additive or synergistic effects but should be formulated in a controlled

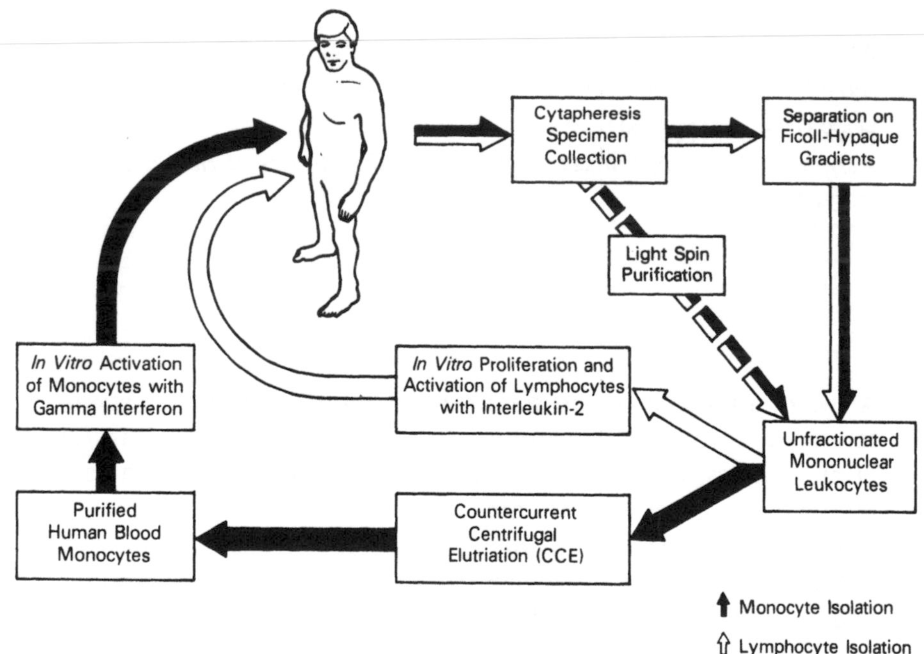

Figure 1 Schematic illustrating the central role of cytapheresis in the lymphocyte and monocyte isolation components of the LAK-IL2 and AKM therapies respectively.

fashion. (2) Effector cells should be purified in a manner suitable for human use. Thus, cell purification strategies that free the cells of pyrogens, pathogens, and toxins are required. (3) Since ACI is dependent upon obtaining sufficient numbers of effector cells to elicit a clinical response, collection procedures that harvest maximal numbers of leukocytes should be identified. For both the LAK and AKM trials, the initial cell isolation procedure currently performed is cytapheresis as shown in Figure 1.

BASIC CYTAPHERESIS TECHNOLOGY

The ability of a cytapheresis cell separator apparatus to efficiently remove a specific component of blood, such as mononuclear cells, is determined by the instrument's efficiency in separating blood into distinctive layers by centrifugation on the basis of differences in cell density (displacement) or drag coefficients in a current (elutriation). The apparatus must then specifically remove the desired cell type, while returning the other blood constituents to the patient/donor. When centrifugal force is applied to whole blood, blood constituents separate into layers according to their respective buoyant densities (displacement). The degree of separation is dependent on the intensity of the centrifugal force applied, the time of centrifugation, and the design of the rotating separation chamber. With the appropriate centrifugal force, the layered components that can be separated according to decreasing specific gravities include: mature red blood cells, young red blood cells (neocytes), granulocytes, mononuclear cells (lymphocytes, monocytes, stem cells), platelets, and plasma (1). Once the distinctive layers are formed, further mechanical principles (such as aspiration and decantation) may be employed to specifically remove the desired components. Blood cell components can also be separated by elutriation, since many of these cellular elements have distinct drag coefficients when suspended by centrifugation in a counterflow media stream; those cells with higher drag coefficients are the first to exit the separation chamber. Since most blood cell separator machines employ elements of both separation principles (displacement and elutriation), calculation of pertinent cell separation parameters may be complex. Each cytapheresis instrument's optimal collection parameters for any specific blood component is determined by the interaction of the separation and removal characteristics of the particular instrument.

Commercial cytapheresis instruments fall into two categories: the intermittent (or discontinuous) flow centrifuge and the continuous flow centrifuge. In the first, blood is combined with an anticoagulant in a fixed ratio before being pumped through a rotating seal into the separation chamber. While in the chamber, less buoyant cellular components are displaced at the periphery of the chamber and the lighter weight cellular components are continuously directed toward an outflow tract at the center of the separation chamber. Sequential separation of blood components (plasma, platelets, mononuclear cells and granulocytes) is achieved by further addition of whole blood to a full centrifuge chamber. The desired blood component can be collected into a separate container by observing the sequential outflow of components and the opening and closing of appropriate clamps. The separation procedure is stopped as soon as the red cells reach the outflow tract; the forward pump is reversed to deviate the uncollected elements into a separate collection bag for reinfusion back to the donor/patient. Examples of present day intermittent cytapheresis machines are the Haemonetics model 30 and Haemonetics V50 (Haemonetics Corporation, Braintree, MA). The former is a semiautomated system that needs constant operator attention while the latter has been further automated to perform the desired procedure, using optical sensors to detect the red cell interface, solid-state chip technology for timing, opening and closing clamps, and operating a variety of donor safety features.

In the continuous flow centrifuges, anticoagulated blood from the donor/patient is pumped via a rotating seal or sealless connection into the separation chamber. The separation of blood cell components takes place in the separation path: vertical chamber systems (Aminco Celltrifuge I and Celltrifuge II Americal Instrument Company, Silver Spring, MD and Fenwal Laboratories, Deerfield, IL) and rigid circular plastic pathway systems (IBM

2997 and Cobe Spectra Laboratories, Inc., Lakewood, CO) exist. The desired component is collected continuously, and other components returned to the donor with a second venipuncture site. In contrast to the Celltrifuge I, the IBM 2997 cannot collect different blood components simultaneously from different exit ports unless another pump and a different separation chamber are used.

The CS-3000 blood cell separator (Fenwal Laboratories, Deerfield, IL) replaces the rotating seal with a multilumen sealless connection. Two chambers are situated in the centrifuge rotor. The separation chamber collects the anticoagulated blood from the donor and separates the red cells and leukocyte- or platelet-rich plasma components. The leukocyte-rich plasma is pumped into the collection chamber, in which the leukocyte component that is needed is concentrated. The use of chamber inserts allows for modification of the configuration of the collection chamber and this permits selective concentrations of cellular elements. The red cells exit from the separation chamber and return to the donor/patient. Cells in the collection chamber remain in the centrifugal field for the duration of the procedure and are not visible to the operator.

COMPARISON OF CYTAPHERESIS MACHINES UTILIZED IN ACI

At present, mononuclear cells (MNC) have been collected with four cell collection instruments: Haemonetics V50, Cobe 2997, Fenwal Celltrifuge II, and Fenwal CS 3000. The Haemonetics V50 can be used to collect, purify, wash, and harvest cells for adoptive cellular immunotherapy (ACI) (2). While the Fenwal Celltrifuge II and Cobe 2997 can only be used to collect the MNC, if further purification is desired, another instrument, such as the Cobe 2991 or manual technique, can be used (3). The Fenwal CS 3000 is also capable of harvesting cells for ACI. Another newly developed instrument with potential ACI applicability is the new automated version of the Cobe 2997, the Cobe Spectra, though little data is available regarding its efficacy.

Table 1 Metaanalytical Comparison of Mononuclear Leukocyte Apheresis Techniques in Normal Donors

Model configuration–(manufacturer)	Celltrifuge I Dispo Bowl–Aminco	Celltrifuge II Dispo Bowl–Fenwal	Model 2990 Reuse Bowl–IBM	Model 2997 Single Stage–IBM/Cobe	Model 30 225 ml–Haemonetics	Model V50 225 ml Bowl–Haemonetics	Model V50 225 ml Bowl Auto Surge–Haemonetics	CS3000 Granulo #3-75–Fenwal[a]	CS3000 Granulo #1-100–Fenwal[b]
Number of accesses	2	2	2	2	1	1 or 2	1 or 2	1 or 2	1 or 2
System volume, ml	335	345	335	198	275	275	275	453	453
Collection volume, ml	200–600	200–600	200–600	200–600	300–400	300–400	300–400	200	200
Maximum extracorporeal volume (ml)[c]	935	945	935	798	745	745	745	453	453
Packed RBCs at risk[d]	110	110	110	65	170	170	170	98	98
Controlled by:	Operator	Operator	Operator	Operator	Operator	Operator/ Computer	Operator/ Computer	Computer	Computer
Principle of separation[e]	Displace	Displace	Displace	Displace	Displace	Displace (30 ml into RBC)	Elutriate	Elutriate	Elutriate
Closed system	No	No	No	No	No	No	No	Yes	Yes

Sealless fluid path	No	Yes	No	No	No	No	No	Yes	Yes
Efficiency of lymphocyte collection	20–60%	20–60%	20–60%	75–85%	90–100	90–100	53–63	65–75	75–85
Granuocytes in collection[f]	5–20%	5–20%	5–20%	0–3%	2–8%	10–20	0–2	0–15	0–3
Efficiency of monocyte collection[f]	NA[g]	40–60%	NA	25–35	90–100	90–100	25–35	55–65	35–45
Required draw rate ml/min	15–80	15–80	15–80	15–70	60–80	60–80	60–80	15–60	15–60
Minutes to process one liter	20	20	20	17	27	27	27	22	18

[a]"Monocyte collection" setting

[b]"Lymphocyte collection" setting

[c]Extracorporeal volume = system volume and collection volume. In the Haemonetics Model 30 and 50 the extracorporeal volume is dependent on patient hematocrit (Hct of 40% employed).

[d]Packed RBC that would be discarded if entire procedure had to be discontinued.

[e]Displacement vs. elutriation as dominant cell separation principle.

[f]Collection efficiency = $\frac{(200) \times (\text{Collection Bag WBC} \times \text{Bag subset \%} \times \text{Bag volume})}{(\text{Preperipheral blood WBC} \times \text{subset \%} \times \text{volume processed}) + (\text{Postperipheral blood WBC} \times \text{subset \%} \times \text{volume processed})}$

[g]NA denotes data not available.

Comparing the efficiencies of collection of these distinct machine types in the ACI patient setting has not yet been accomplished. However, the collection efficiencies of these machines have been studied in some detail in the normal volunteer setting (1–9). As summarized in the meta-analysis shown in Table 1, both the continuous and discontinuous cytapheresis machines can collect substantial numbers of mononuclear leukocytes from normal volunteers; collection efficiencies must be higher in the discontinuous models to compensate for lower overall volume of blood processed. Of course, the settings for patients are different, particularly in those patients who have been pretreated with biologicals such as interleukin-2 (IL-2). Thus, we must still await controlled data from the ACI patient setting before we can conclude which machine type produces the highest volume and quality of the LAK and AKM leukocyte substrates.

CYTAPHERESIS IN ACI TRIALS

The quality and quantity of mononuclear leukocytes collected by cytapheresis depends on several factors: (1) total volume of blood processed; (2) the concentration of peripheral blood leukocytes; (3) centrifugal force and rate of blood flow acceleration; (4) the anticoagulant to whole blood ratio; (5) the displacement of blood elements into the white blood cell interface (buoyant penetration); (6) the mobilization of leukocyte reservoirs during the procedure; and (7) finally, as-yet undetermined patient-to-patient variables (4). The patient's peripheral leukocyte count, volume of blood that can be processed and centrifugal force and flow rate of the pheresis instrument are critical variables affecting the number of cells collected at the end of the procedure. The patient tolerance and the risk of toxicity from blood additives (such as anticoagulants) limit the length of the procedure (practical limit per procedure is 4 h), while the risk of inducing anemia limits the number and frequency and (to a certain extent) the length of the procedure. Because the functional status of the isolated cells is critical, strategies are being developed for collection of the target number of mononuclear leukocyte subset cells required for clinical trials in the shortest possible time.

The concentration of mononuclear leukocyte subsets in the peripheral blood is the factor that most directly affects the number of subset cells collected by cytapheresis. A high precollection peripheral blood lymphocyte count yields the most lymphocytes, whereas a high precollection number of human blood monocytes best predicts the effectiveness of the cytapheresis collection of this cell type (5). To date, no detailed analysis of predictive factors for the efficiency of collection of lymphocyte subsets (T, B, and NK cells) have been performed. Sedimentation velocity profoundly affects the efficiency of leukocyte subsets separation. Hester et al., using a Cobe 2997 have observed that (1) lymphocyte separation was significantly increased with centrifugal force from 580 rpm to 640 rpm and (2) a plateau was reached at 640 to 1160 rpm range (4). Similarly, Wright et al. (3) found that at higher centrifugal speeds, more lymphocytes and monocytes were collected per procedure. However, at speeds approaching 1500 rpm, increasing platelet clumping was observed. It is likely that the current centrifugal collection cytapheresis techniques cannot be manipulated to enrich for subsets unless the patient is pretreated to increase the number of circulating subset cells.

Another factor which can affect the number and type of mononuclear cells collected is their displacement through the other blood components. In normal donors, it appears that lymphocyte buoyant density places them relatively close to the plasma interface with most cytapheresis rotor systems, such that maximal numbers of lymphocytes are collected at relatively low hemoglobin concentrations (4 g/dl or less) (4,8). In contrast, the displacement characteristics of human blood monocytes appear to be substantially different. A large number of monocytes can be collected at an interface hemoglobin concentration of 10 g/dl when compared to a hemoglobin concentration of 1 g/dl (8,9). In cancer patients pretreated with IL-2, it has been observed that the best quality final unfractionated mononuclear leukocyte product is obtained when the collection is at a hemoglobin concentration between .5 and 1.0 g/dl. This observation derives in part from the fact that the patients receiving bolus rIL-2 may demonstrate a striking peripheral blood eosinophilia; when present, these eosinophils can be very difficult to separate from mononuclear cells by Ficoll-Hypaque gradients. The leukocyte concentrate collected at the .5–1.0 g/dl hemoglobin range allows for optimal separation of effector cells with subsequent Ficoll-Hypaque gradient centrifugation when employed. In addition, because LAK cell patients undergo cytapheresis daily for 5 days, this collection strategy minimizes the transfusion requirement of these patients. In contrast, the collection of monocytes for the AKM immunotherapy generally requires higher hemoglobin concentrations in the collection bag as previously described for normal donors (9) and the procedure is generally performed weekly.

Mobilization of leukocyte subsets from extravascular sites during cytapheresis probably contributes to the donor-to-donor variability of leukocyte collections. Two observations by Hester et al. (4) suggest that mononuclear leukocyte mobilization occurs during cytapheresis: (1) there is a nonlinear fall in peripheral blood mononuclear cell concentration during the procedure, and (2) the residual number of circulating mononuclear cells at the end of the procedure is greater than expected considering the number of mononuclear cells harvested. These findings indicate that sufficient mononuclear cell reservoirs exist in humans to reconstitute acute cell losses, at least transiently. In addition, the ability of donors to increase production of leukocytes within the first few days of depletion must be substantial (4,5). To date, very little work has been done to determine which mononuclear cell subsets replenish the circulation following acute and chronic mononuclear cell depletion procedures. The patient's disease and administration of biological response modifiers further complicate the scientific analysis of this issue.

Finally, the variability between normal cytapheresis donors, even when all measured preprocedure variables (such as normal history and physical examinations, absence of drug ingestion, normal chemistry profiles, normal sedimentation rates, and normal complete blood counts) are controlled, presents a significant impediment to obtaining reproducible results from mononuclear cell collections. In an anlaysis of 99 mononuclear cell procedures in 36 normal donors, "high yield" lymphocyte donors and "low yield" lymphocyte donors were detected. However, lymphocyte collection yields varied substantially from day to day even in collections from the same donor. In contrast, no evidence for consistently "high" or "low" monocyte donors was found. There was no obvious explanation for the considerable variability of monocyte and lymphocyte collections from individual donors or for the overall donor pool (9).

The influence of the cell separation instrument used in ACI trials is an important technical aspect that needs to be evaluated. Several factors other than purity and the number of cells collected per procedure, should be considered: (1) minimum time necessary to harvest the desired number of leukocyte subsets; (2) the potential advantage(s) of single-needle versus double-needle techniques; (3) the amount of hemoglobin removed per procedure; and (4) most importantly, the evaluation of the function of the leukocyte subsets harvested by different cytapheresis techniques.

During clinical trials, several additional complications of cell collection have become evident. In the LAK + IL-2 protocol, cancer patients are routinely pretreated with 100,000 μg/kg of recombinant interleukin-2 (rIL-2) every 8 h for 3 to 5 days followed by a two-day respite prior to performing cytapheresis. It is desirable to discontinue rIL-2 administration before beginning cytapheresis in order to increase the yield of LAK cells, because LAK precursor cells rapidly disappear from the peripheral blood during rIL-2 administration; there is a marked rebound of LAK cell precursors in the peripheral blood after discontinuation of rIL-2, such that maximal number can be obtained during initiation of cytapheresis (11,12). As a result, 10 to 50×10^9 mononuclear cells are obtained from each collection procedure. A single course of therapy generally consists of one week of cytapheresis (13,14) which yields 1 to 5×10^{11} cells per patient per course of therapy. Similar strategies are being devised for harvesting human blood monocytes for AKM therapy. In view of present evidence indicating that colony-stimulating factor 1 (CSF1) is a potent stimulator of monocyte release into the peripheral blood, this BRM is envisioned to be a potential priming biologic for the AKM immunotherapy (15). The exact effect of these biological priming steps in the sedimentation and drag coefficient characteristics of specific target leukocyte subsets has not yet been well defined but is the subject of intense investigation.

NEW STREAMLINED LAK PRECURSOR HARVESTING TECHNIQUES

Instrumentation research and development efforts are underway to modify the cytapheresis procedures and allow for more precise collection of the target mononuclear cells. Strategies to accomplish this goal include automated Ficoll-Hypaque (F/H) purification using the Fenwal CS 3000 blood separator (6,10). F/H purification gradients during the collection phase are used to obtain a more purified mononuclear leukocyte concentration (and to avoid subsequent F/H gradient centrifugation in the laboratory). The MNCs collected by this approach have very little granulocyte contamination.

LAK precursor collection can be further simplified by replacing the F/H purification with a light centrifuge spin (LS) to remove excess platelet contamination (16,17). The Mayo Clinic undertook a study comparing the mononuclear leukocyte (MNL) recovery, platelet and red cell removal, purification time, in vitro LAK cytotoxicity, and phenotypic characterization of the lymphocytes collected by F/H purification versus LS purification, using the CS 3000. Six donors were leukapheresed at 60 ml/min for 90 min with the CS 3000 using the LAK cell collection procedure. Following a LS (3 min at 1000 rpm), the MNL collection was divided into two equal aliquots, one aliquot purified by F/H and the other aliquot purified by LS. Both aliquots were then washed, cultured for 5 days in IL-2, and harvested with the CS 3000. Cell counts were performed on the Coulter S+4. LAK activity was measured using a 4 h ^{51}Cr release assay. Lymphocyte phenotype was determined by flow cytometric analysis using the FACS IV Analyser flow cytometer. Postpurification aliquots for F/H and LS averaged

1.7×10^9 MNC and 2.4×10^9 MNC, respectively. Additional findings suggesting superiority of LS purification included: (a) postpurification MNL recovery (89% vs. 60%), (b) total MNL recovery (64% vs. 40%), (c) purification time (54 min vs. 129 min), and (d) LAK cell generation by phenotype analysis. After culture, purification was similar to F/H purification in: (a) platelet removal, and (b) in vitro LAK cytotoxicity. F/H was superior to LS in red cell removal (77% vs. 19%). LS was found to be superior to F/H when considering efficiency, economics, simplicity, and LAK cell generation, and comparable in platelet removal and in vitro LAK cytotoxicity.

CONCLUSIONS

Technical factors associated with the collection and purification of cytotoxic effector cells are important for promoting widespread clinical testing of ACI. Many of the complex laboratory manipulations of LAK and AKM cells have already been simplified, such as the development of a suitable automated closed system of leukocyte collection, serum-free media, and nonadherent laboratoryware. Further automation of these steps is required and a rapid standardized format for comprehensive quality control needs to be developed. An automated cytapheresis closed-system collection device will hopefully be developed that will allow for the purification of leukocyte subsets directly at the bedside. Not only will this development promote a more standardized final ACI product, the elimination of extraneous (and possibly antagonistic) blood cellular components may minimize the laboratory resources required for generating the desired leukocyte subset activation in vitro.

REFERENCES

1. Huestis D W, Bone J R, Busch S. Haemapheresis. In Practical Blood Transfusion, 3rd edition. Little Brown, Boston, 1981.
2. Bell A J, Ubrichtham M, Stevenson F K, Hamblin T J. Genera-

tion of lymphokine activated killer cells in a totally closed system. Plasma Ther Transfus Technol 1987; 18:371–376.
3. Beaujean F, Gourdin M F, Farcet J P, et al. Separation of large quantities of mononuclear cells from human blood using a blood processor. Transfusion 1985; 75:152–154.
4. Hester J P, Kellog R M, Freireich E J. Mononuclear-cell (MNC) collection by continuous-flow centrifugation (CFC). J Clin Apheresis 1983; 1:197–207.
5. Stevenson H C, Beman J, Riggs C, Kanapa D J, Miller P. Multivariate analysis of factors associated with monocyte and lymphocyte collection by cytapheresis of normal donors. Plasma Ther Transfus Technol 1984; 5:335–344.
6. AuBuchon J P, Carter C S, Adde M A, Meyer D R, Klein H G. Optimization of parameters for maximization of platelet pheresis and lymphocytapheresis yields on the Haemonetics Model V50. J Clin Apheresis 1986; 3(2):103–110.
7. Dowling R, Weber V, Osborne L, Klein H G. Mononuclear cell collection using various techniques. J Clin Apheresis 1984; 2:32–40.
8. Wright D G, Karsh J, Fauci A S, Klippel J H, Decker J L, O'Donnell J F, Deisserroth A B: Lymphocyte depletion and immunosuppression with repeated leukapheresis by continuous flow centrifugation. Blood 1981; 3:451–458.
9. Stevenson H C, Beman J, Huffer T L, Riggs C, Kanapa D J, Fer M, Miller P. Differential sedimentation characteristics of human monocytes and lymphocytes during cytapheresis. Plasma Ther Transfus Technol 1984; 5:323–334.
10. West W H, Tauer K W, Yannelli J R, et al. Constant infusion recombinant IL-2 in adoptive immunotherapy of advanced cancer. N Engl J Med 1987; 316:898–906.
11. Lotze M T, Matory Y L, Ettinghausen S E, Rayner A A, Sharrow S O, Seipp C Y, Custer M C, Rosenberg S A. *In vivo* administration of purified human IL-2. J Immunol 1985; 135: 2865–2872.
12. Beckner S K, Urba W J, Clark J, Steis R G, Longo D L. LAK cells: Optimization of *ex vivo* conditions for clinical trials; 1987 (submitted).

13. Rosenberg S A, Lotze M T, Muul L M, Leitman S, Chang A E. Ettinghausen S E, Matory Y L, Skibber J M, Shiloni E, Vetto J T, Seipp C A, Simpson C, Reichert C M. Observations on the systemic administration of autologous lymphokine activated killer cells in recombinant IL-2 to patients with metastatic cancer. N Engl J Med 1985;313:1485–1492.
14. Rosenberg S A, Lotze M T, Muul L M, et al. Progress report on the treatment of 157 patients with advanced cancer using lymphokine activated killer cells and IL-2 or high dose IL-2 alone. N Engl J Med 1987;316:889–897.
15. Warren M K, Ralph P. Macrophage growth factor (CSF-1) stimulates human monocyte production of interferon, tumor necrosis factor and colony stimulating activity. J Immunol 1986;137:2281–2290.
16. Burgstaler E, van Haelst C, Pineda A A, et al. Simplification of LAK cell generation: Ficoll-hypaque versus light spin purification. J Clin Apheresis (submitted).
17. Abersold P, Carter C, Hyatt C, et al. A simplified procedure for generation of human lymphokine-activated killer cells for use in clinical trials. J Immunol Methods 1988; 112:1–7.

12

Clinical/Technical Challenges in Adoptive Cellular Immunotherapy: In Vitro Culture Techniques

JOHN R. YANNELLI and MARTIN R. JADUS
Biotherapeutics, Inc., Franklin, Tennessee

SUZANNE BECKNER and LEOCADIO V. LACERNA
National Cancer Institute, Frederick, Maryland

ROBERT K. OLDHAM
Biological Therapy Institute, Franklin, Tennessee

Adoptive cellular immunotherapy (ACI) refers to the passive transfer of immune reactive cells such as NK cells, T lymphocytes, or monocytes, into tumor-bearing hosts to mediate antitumor responses by either direct (cell-to-cell contact) or indirect (soluble factor) mechanisms. Much effort is currently underway in the area of treating cancer patients with these novel regimens. Such approaches now include three separate clinical trials being conducted at various centers across the country. These trials include lymphokine-activated killer (LAK) cells, interferon-γ (IFN-γ) activated killer monocytes (AKM), and tumor-derived activated cells (TDAC), or tumor-infiltrating lymphocytes (TIL). Clinical progress in this area has been well summarized in Chapters 3 to 6 and 8.

In this chapter, we will focus our discussion on the strategies that have been developed for the in vitro handling and preparation of the immunological effector cells used in ongoing clinical trials being conducted by the National Biotherapy Study Group (NBSG) and elsewhere. In the LAK-type biotherapy, mononuclear leukocytes (lymphocytes and monocytes) are removed from the body of the cancer patient and activated in the presence of a high concentration of recombinant human interleukin-2 (rIL-2) (100–1000 units/ml) to become nonspecific cytotoxic cells. After this antitumor activity is upregulated in vitro, the LAK cells are reinfused back into the cancer patient. In the AKM therapy, the baseline cytotoxic activity of human blood monocytes is upregulated with IFN-γ, transforming these nonspecific cytotoxic cells into AKM.

These ACI protocols are representative of the types of laboratory challenges faced daily for activating cells in vitro and returning them in a pharmacologically acceptable fashion to cancer patients, and they raise a number of issues that are not normally faced by researchers doing standard cell culture. Specifically, the challenge of the LAK cell protocol revolves around processing the large numbers of cells routinely handled (3.0×10^{10} to 2.0×10^{11}) and the requirement to maintain these cells as a pharmaceutical-type reagent. On the other hand, the AKM protocol laboratory staff must successfully process leukocytes that are very fastidious and readily clump to standard laboratory plasticware. Methods have been developed (see Chap. 11) for collecting cells (cytapheresis) from the patients with a minimal amount of manipulation. In addition, an automated system of cell culture (1) and harvesting (2,3) is currently being used. Thus, the principal goal of ACI cell-processing laboratories is to process as many cytotoxic precursor cells from the patient as possible. Once the cells arrive in the laboratory, the cells must be cultured in such a way as to provide the largest number of clean (microorganism- and endotoxin-free), activated (cytotoxic) cells ready for use in the cancer patient. These are the points to be discussed in the remainder of this chapter.

IN VITRO PROCESSING OF LAK CELLS

Mononuclear cells obtained from the cytapheresis of cancer patients or normal donors must undergo in vitro cell culture in high-dose IL-2 to display maximal LAK cytolytic activity against tumor targets. Interestingly, the question is why the LAK precursor cells must be activated "ex vivo" when the patients have already received high-dose IL-2? Our current thinking suggests that the purpose of the IL-2 infusions is to cause the proliferation and mobilization of LAK precursors. When the IL-2 infusion is stopped, the LAK precursors traffic to the peripheral circulation. However, the concentration of IL-2 achieved in vivo does not appear to be high enough to result in significant cytotoxic activity of the cells. Thus, the LAK precursors must be cultured in a dose of IL-2 that cannot be safely achieved in vivo. In vitro cytolytic activity against Daudi (a LAK-sensitive tumor cell) appears as early as 24 hr and gradually increases over the course of 3 to 7 days of culture (Table 1). Preliminary results from our laboratory suggest that as little as 15 min might be required for the delivery of the signal to the LAK precursor to become cytotoxic. The in vitro cell culture may only be required for the cell to mature and to further develop its cytotoxic potential (Horton and Yannelli, unpublished results).

A number of laboratories have characterized IL-2-activated mononuclear cell populations using flow cytometry to identify the LAK effector cell (4–6). In our studies, most of the cells obtained from the cytapheresis procedure are T cells ($CD4^+$ and $CD8^+$) with a variable percentage of NK cells (Table 2). Only a small percentage of cells express B-cell or macrophage markers. After 3 to 5 days of cell culture, the cells whose percentage is most strikingly increased are the NK cells. In addition, these cells begin to express the T-cell activation antigen, TA1. Interestingly, cytotoxic activity of the NK cells in primary LAK cell cultures appears to correlate with the appearance of this TA1 marker (7).

Because the LAK cell appears to be characterized by an NK phenotype, it is curious that more effort has not been made to

Table 1 Kinetics of LAK Appearance in Normal Donors (A-D)

	Lytic units[a] per 10^7 effectors							
	K562 target[b]				Daudi target			
Day	A	B	C	D	A	B	C	D
0	40	34	47	20	41	13	0	16
1	34	161	384	180	104	153	223	95
2	243	417	434	524	712	973	1939	1166
3	334	1021	1859	854	1259	868	1980	2028
4	968	ND[c]	ND	ND	1070	ND	ND	ND
5	ND	1068	2747	2062	ND	2065	2060	2300

[a]The number of lytic units is obtained by determining the number of effector cells necessary to give 33% lysis of 2.5×10^3 target cells in a 4-h assay and dividing that number into 1×10^7 effector cells.
[b]Tumor targets used were the NK-sensitive human erythroleukemia cell line K562, and the LAK-sensitive human lymphoblastoid cell line Daudi.
[c]Not determined.

purify the NK cells and grow them as a single population for the cancer patients. The rationale may be twofold: (a) We are not sure of the contribution of the "irrelevant" cells (IL-2-activated T cells, B cells, and macrophages) to the in vivo LAK phenomenon. These cells can release large quantities of other lymphokines [IFN-α, IFN-γ, tumor necrosis factor (TNF; Dupere and O'Connor, Biotherapeutics, unpublished results)], which may have antitumor effects; and (b) NK cells, programmed to become LAK cells, although generally observed in G_2 or M phase following IL-2 activation, may not be capable of a crucial number of cell divisions. Thus, the number of LAK effector cells available for therapeutic purposes may be minimal if this approach were attempted.

In Biotherapeutics LAK–IL-2 clinical studies, the patients receive continuous infusion IL-2 for 4 to 5 days followed by a 24 to 48 hr rest (8; refer to Chap. 5). Interestingly, during the IL-2 infusions, the patients experience a dramatic leukopenia (specifically lymphopenia). However, when the IL-2 infusion is discontinued,

Table 2 Flow Cytometric Analysis[a] of a Cancer Patient's Mononuclear Cells Before and After Incubation with IL-2

	% Mononuclear cells expressing:					
Procedure[b]	T4	T8	B1	MO2	NKH	NKH/TA1[c]
Cytapheresis	15	26	3	4	35	1
Reinfusion	13	64	9	2	48	34
	Lytic units[d]/10^7 effector cells against:					
	K562					Daudi
Cytapheresis	2086					87
Reinfusion	3688					3758

[a]Performed using direct immunofluorescence detected with an Epics V Flow Cytometer (Coulter Electronics, Hialeah, FL.
[b]Cytapheresis material was tested before culture in IL-2. Reinfusion refers to LAK cells cultured for 3 days in IL-2.
[c]% of cells coexpressing the NK marker NKH and the T-cell activation marker, TA1.
[d]Lytic units defined in footnote of Table 3.

there is a rebound of leukocytes which includes an increased percentage of lymphocytes (lymphocytosis) into the peripheral blood. We then begin the cytapheresis of the cancer patients. Seven to 12 liters of peripheral blood are processed at a flow rate of 50 ml/min. The final product, which contains predominantly lymphocytes and monocytes, is washed free of residual platelets. The platelets are returned to the cancer patient. The patient is cytapheresed in this manner for 4 consecutive days. Table 3 shows the total white blood cell (WBC) counts and Wright stain differential analysis of a representative cancer patient on LAK–IL-2 therapy. The WBC counts are elevated over normal during the course of 4 days after the halt of IL-2 infusions. However, the WBC counts generally drop following each cytapheresis, the percentage of lymphocytes generally decreases in the peripheral blood, whereas the percentage of granulocytes increases.

Table 3 Peripheral Blood Profile of LAK Patients Before and After Cytapheresis on a Fenwal CS3000

Procedure[a]	WBC × 10^{-3}	% Lymphocytes[b]	% Atypical lymphocytes	% Monocytes	% Neutrophils	% Eosinophils
C1						
Precount	15.4	50	11	2	35	2
Postcount	10.0	29	2	3	58	2
C2						
Precount	11.7	33	18	9	36	4
Postcount	9.2	27	9	5	56	3
C3						
Precount	10.4	30	19	4	43	4
Postcount	7.3	30	5	6	55	4
C4						
Precount	10.7	19	14	5	54	8
Postcount	7.3	10	3	5	81	1

[a]Procedure refers to sequential (1 per day) cytapheresis. Precounts refer to WBC count and differential done on samples drawn less than 4 hours before the cytapheresis. Postcounts were done on samples drawn less than 1 hour after the cytapheresis.
[b]Wright stain differential analysis was done using a Leukostat stain kit obtained from Fisher Scientific (Fairlawn, NJ).

Wright stain differential data pooled from 100 different cytapheresis procedures of 25 separate cancer patients participating in a Biotherapeutic LAK–IL-2 study are presented in Table 4. The cytapheresis product obtained by the CS3000 contains predominantly mononuclear cells, lymphocytes, and monocytes. Granulocytes comprise less than 5% of the final product. The hematocrit is generally less than 3% (represented by an RBC/WBC ratio of <3:1). In addition, the platelet/WBC ratio is in the range of 10:1 or less. Based on these data, we decided to eliminate the Ficoll–Hypaque isolation which removes red blood cells, granulocytes, and platelets from mononuclear cell preparations. Previously, we had developed a method for Ficoll–Hypaque isolation using the Fenwal CS3000 (1). Our recoveries using a manual method of isolation were 50% ± 23%, and were improved by the automated methodology to 65% ± 26%. However, because the cytapheresis product did not contain large numbers of contaminating RBCs and granulocytes, elimination of this step resulted in increased numbers of cells available for cell cultures. In addition, we showed that by not subjecting the cells to this purification (a) the cancer patients received more cells after cell culture, and (b) the LAK cells

Table 4 Wright Stain[a] Differential Analysis of Cancer Patients Leukapheresed Blood Before and After Ficoll-Hypaque Purification

	Percentage[b] of total cells			
	Granulocytes	Lymphocytes	Monocytes	RBC:WBC[c]
Before purification	3.2 ± 4.1	86.4 ± 6.7	8.2 ± 6.7	17:1
After purification	1.5 ± 1.1	89.2 ± 4.1	5.1 ± 2.3	1:3

[a]Wright stain differential analysis were done using a Leukostat stain kit obtained from Fisher Scientific (Fair Lawn, NJ).
[b]At least 100 cells were evaluated per sample. The values presented are the mean and the standard deviation of 24 samples obtained from 8 different cancer patients.
[c]Red blood cell to white blood cell ratio.

had more cytolytic activity (Fig. 1) (9). Thus, most LAK–IL-2 centers have now removed this procedure, which was costly in terms of time, money, cell product, and cytolytic potential.

The cytapheresis product enters the laboratory in a volume of 200 ml of saline. The mean yield of cells per cytapheresis is $3.86 \times 10^{10} \pm 2.3 \times 10^{10}$. Considering the large numbers of cells collected from the patients on a daily basis, we saw the need for improvements in large-scale culture capacity. Historically, cells were cultured at low cell density ($1.0–1.5 \times 10^6$ cells/ml) in 3-L roller bottles (10). Recently, an automated system of cell culturing that uses 1- or 3-L capacity Fenwal platelet storage bags (PL732 plastic) has been developed (Fig. 2). These bags are permeable to gases (CO_2 and O_2) but not to liquids. In comparative studies, these bags were equivalent to culturing in plastic T175-cm^2 flasks in terms of cell recoveries, cell viability, cell surface phenotype, and ability to lyse tumor targets (1,2). In addition, we were able to increase the culture cell density to 3.0×10^6/ml using these bags. Recently we have determined that culture bags made of Teflon can also be used for LAK cell culture (Yannelli et al., manuscript submitted). The advantage of using such bags (PL732 or Teflon) is that they can be utilized as a closed system in combination with standard sterile-tubing kits. The bags are filled using a Fenwal solution transfer pump and associated tubing. Currently, the flow rate using this system is 300 ml/min, thus, a single bag takes around 3 min to fill (3-L bag filled with 1-L volume). The bags are then incubated at 5% CO2 in air at 37°C for 3 to 5 days.

After 3 to 5 days of cell culture, the CS3000 is utilized as a cell-harvesting machine. With a standard open-tubing kit, the machine is first primed in an automated mode to remove air from the lines. Bags of cells are then connected to the machine using sterile-tubing kits. When using both whole-blood and plasma pumps, the flow rate is 176 ml/min. This flow rate is currently the limiting step to decreasing the harvest time. Efforts are currently underway to improve this flow rate, and thus decrease harvest time dramatically. Based on 100 runs, the recovery of cells after the culture

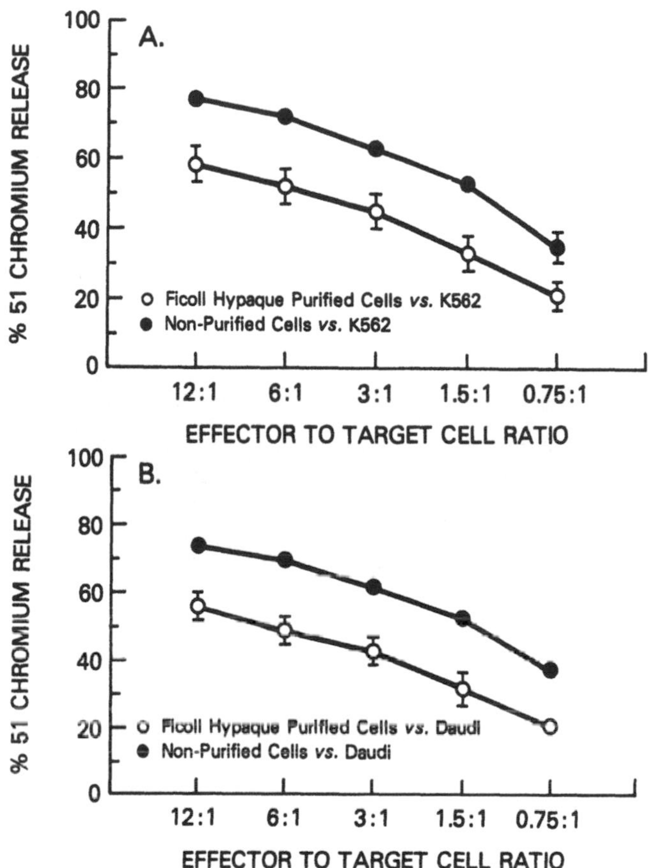

Figure 1 Comparison of the cytolytic activity of Ficoll-Hypaque purified versus nonpurified cancer patient mononuclear cells. The target cells used were the NK-sensitive line K562 in panel A and the LAK-sensitive line Daudi in panel B. The percentage cytotoxicity was determined after 3–5 days of cell culture in IL-2. The mononuclear cells were obtained from different leukapheresis procedures of 10 cancer patients receiving LAK-IL-2 therapy. The data are expressed as mean ± standard error of the mean ($n = 28$).

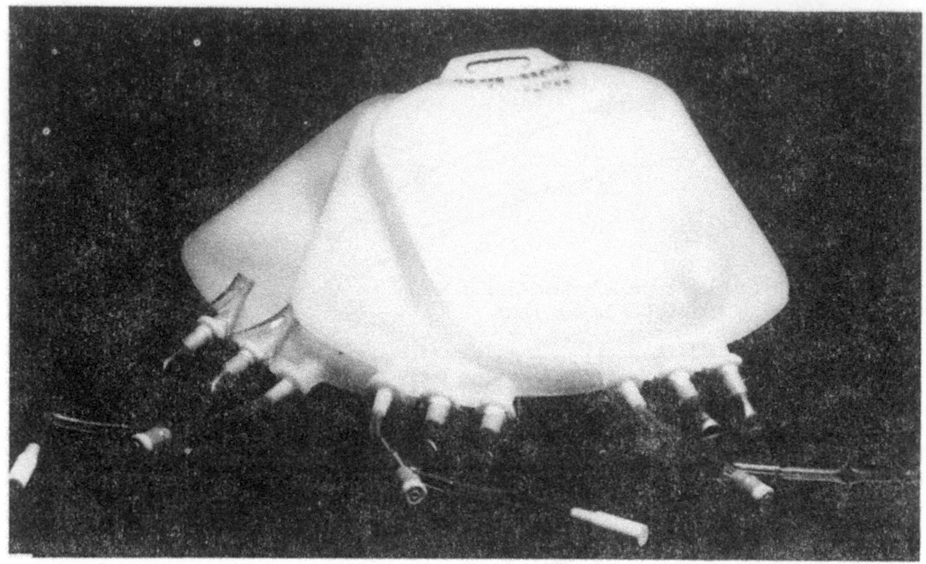

Figure 2 Representative examples of PL732 polypropylene culture bags (Baxter-Travenol, Deerfield, Illinois) used in activation of LAK, TIL/TDAC, and AKM.

and harvesting procedure has been 71% ± 21% cells. The viability of the cells is generally in the range of 90% to 100%.

To monitor the samples for bacterial and fungal growth, three separate tests are done. Samples are routinely drawn 24 hr after the establishment of cell culture and sent to a hospital microbiology laboratory to be cultured in blood culture bottles. This tests detects any microbial growth. In addition, on the day of the harvesting procedure and reinfusion, the final product is tested with both a Gram stain and a limulus amebocyte lysate test (LAL). The Gram stain distinguishes gram-positive or gram-negative organisms. Yeast and fungi can also be visualized in the Gram stain. The LAL determines the level of endotoxin (ET) in the product to be infused. Our endotoxin levels average 0.19 ETU/ml. The cells are transported to the hospital for reinfusion in 5% human albumin plus 1000 units/ml IL-2. Our final reinfusion volume is usually 200 ml, with a cell concentration, on the average, of 100×10^6 cells/ml.

In the original clinical trial of Rosenberg et al. (12), patient LAK cells were activated in RPMI containing 2% human AB serum, a very expensive reagent in short supply. Although RPMI that contained human A serum was shown to be acceptable for the

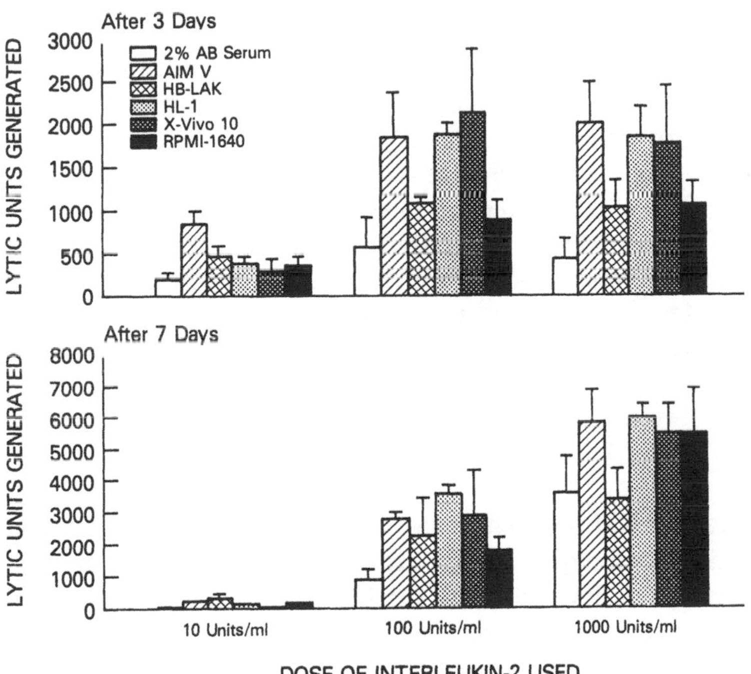

Figure 3 The effect of IL-2 dose upon LAK cell induction in the various serum-free media upon testing on day 3 and day 7. Human LAK cells were cultured for 3 or 7 days in the various media in the presence of 10, 100, or 1000 Cetus units/ml IL-2. The cells were tested in a 4-hr ^{51}Cr-release assay against Daudi target cells. The lytic units were calculated for cells generated in each medium. The results of each medium from four experiments were calculated for the mean lytic unit value along with the standard error.

generation of LAK cells from patients with blood type A and O (13), the availability and variability of human serum, as well as the expense, represents a major limitation to the general application of this particular therapy. Yet another problem with the reliance on human serum is the possibility of disease transmission, such as hepatitis A or HIV-1 infection, after LAK plus IL-2 therapy (14). To avoid this possibility, we determined that serum was not required at all for the generation of human LAK cells (15, 16). This led us to test serum-free medium formulations developed by various companies. The results of these studies are summarized in Fig. 3.

In general, all commercially available serum-free formulations that were tested supported the generation of human LAK cells. However, AIM V (Gibco, BRL, Grand Island, New York), HLI (Ventrex, Portland, Maine), and X-VIVO 10 (M.A. Bioproducts, Walkersville, Maryland) were consistently superior formulations. Culture times could be extended in serum-free formulations to 7 days with increased cytotoxic activity of the LAK cells observed. The routine use of serum-free defined medium provided consistency to LAK for cancer clinical trials and also bypassed problems associated with the use of human serum.

IN VITRO PROCESSING OF AKM CELLS

In the area of AKM research, many of the previously cited LAK in vitro-handling techniques are applicable to AKM; however, techniques have been developed for further purification of human blood monocytes from normal donors and cancer patients. This is a two-step process, consisting of Ficoll–Hypaque gradient separation of the unfractionated mononuclear leukocytes from the cytapheresis procedure followed by countercurrent centrifugal elutriation (CCE) (16). Up to 10^9 monocytes can now be purified daily from cancer patients; this negative selection process allows these vigorously adherent cells to be purified in suspension (Fig. 4). It is important that the monocyte purification techniques employed not be injurious to the activity of the cell and not add toxins or in-

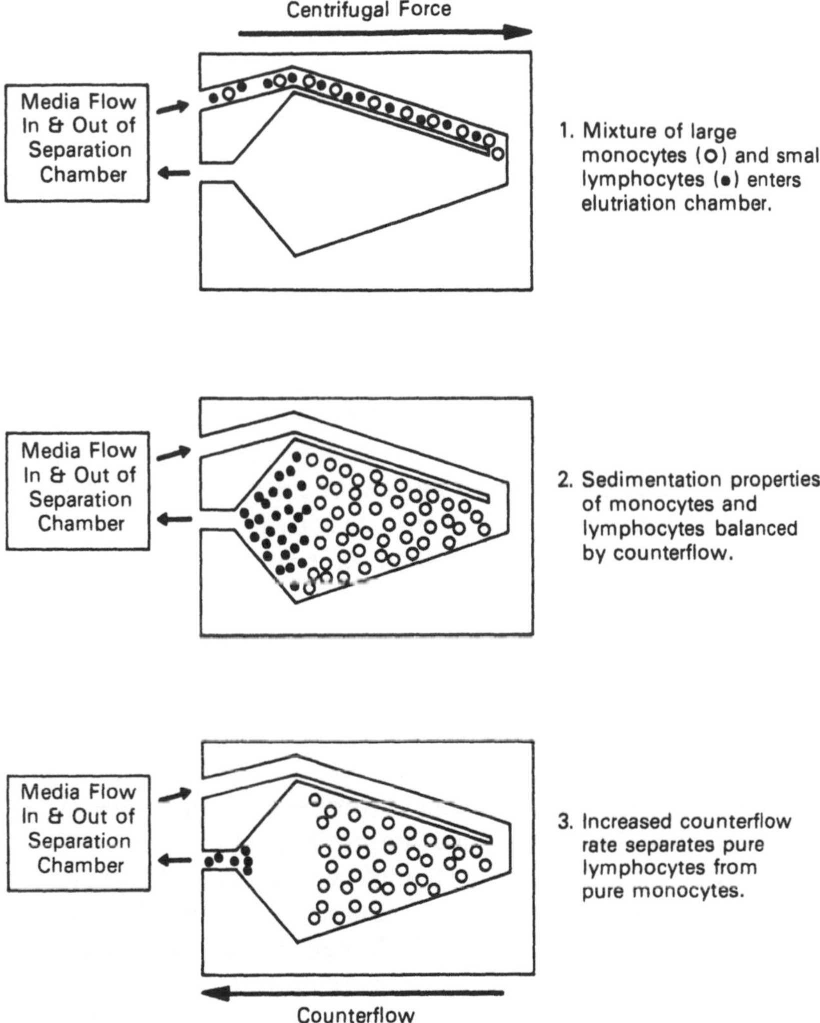

Figure 4 Diagrammatic representation of the countercurrent centrifugal elutriation (CCE) process used to separate monocytes from lymphocytes.

fectious contaminants that might prove injurious to the patient upon reinfusion. The purity and viability of the monocyte preparations obtained are determined by nonspecific esterase straining, Wright's staining, and latex bead ingestion, as previously described (16). In addition, maintenance of the single-cell suspension state of purified human monocytes, utilizing specially developed culture plates and labware, has been achieved that promotes excellent cytotoxic activity of human AKM when cultured for up to 48 hr in rINF-γ (17). This suspension culture technique, utilizing nonadherent polypropylene and Teflon labwear (see Fig. 2), promotes the native characteristics of these cells, avoiding possible alteration or activation by adherence. The AKM are sterilely cultured in the absence of antibiotics. A serum-free medium has been developed for Phase I trials that allows for culture of AKM in suspension (17) and supports the cytotoxic activity of human AKM as determined by a novel 72-hr serum-free suspension cytolytic assay (reviewed in Chap. 7; 18).

CONCLUSIONS

It is clear that many of the complex laboratory manipulations involved in the generation of human cytotoxic cells for clinical ACI trials have been simplified. These improvements include the development of the automated system of cell culture and harvesting, nonadherent labware, as well as suitable serum-free media. Additional automation of these steps is desired, and a standardized format for comprehensive quality control needs to be developed. However, the current LAK culture system has proved broadly applicable in that it is currently utilized in the generation of TDAC cells for clinical trials. This automated system is presently being adapted for the in vitro activation of AKM as well.

REFERENCES

1. Yannelli JR, Thurman GB, Dickerson SG, Mrowca A, Sharp E, Oldham RK. An improved method for the generation of hu-

man lymphokine activated killer cells. J Immunol Methods 1987;100:37.

2. Yannelli JR, Dickerson SG, Sharp E, Thurman GB, West WH, Oldham RK. An automated procedure for the generation and harvesting of human lymphokine activated killer cells (LAK). Proc Am Assoc Cancer Res 1987;28:371.
3. Muul LM, Nason-Burchenal K, Carter CS, Cullis H, Slavin D, Hyatt C, Director EP, Leitman SF, Klein HG, Rosenberg SA. Development of an automated closed system for generation of human lymphokine activated killer cells (LAK) for use in adoptive immunotherapy. J Immunol Methods 1987;101:171.
4. Phillips HJ, Gemlo BT, Myers WW, Rayner AA, Lanier LL. In vivo and in vitro activation of natural killer cells in advanced cancer patients undergoing combined recombinant interleukin-2 and LAK cell therapy. J Clin Oncol 1987;15:1933.
5. Rosenberg SA, Lotze MT. Cancer immunotherapy interleukin-2 and interleukin-2 activated lymphocytes. Annu Rev Immunol 1986;4:681.
6. Tilden AB, Itoh K, Balch CM. Human lymphokine activated killer cells (LAK): Identification of two types of effector cells. J Immunol 1987;138:1068.
7. Yannelli JR, Desch C, Shults K, Houston J, Stelzer G. Characterization of human lymphokine activated killer cells (LAK): Cytotoxicity and cell surface phenotype. Fed Proc 1987;46: 483.
8. West WH, Tauer KW, Yannelli JR, Marshall GD, Thurman GB, Orr D, Oldham RK. Constant infusion of recombinant interleukin-2 in adoptive immunotherapy of advanced cancer. N Engl J Med 1987;316:898.
9. Yannelli JR, Thurman GB, Mrowca-Bastin A, Pennington CS, West WH, Oldham RK. Enhancement of human LAK cell cytolysis and a method for increasing LAK cell yields to cancer patients. Cancer Res 1988;48:5696.
10. Muul LM, Director EP, Hyatt C, Rosenberg SA. Large scale production of human lymphokine activated killer cells for use in adoptive immunotherapy. J Immunol Methods 1986; 88:265.
11. Rosenberg SA, Lotze MT, Muul LM, Leitman S, Chang AE,

Ettinghausen SE, Matory YL, Shiloni JM, Vetto JT, Seipp CA, Simpson C, Reichert CM. Observations on the systemic administration of autologous lymphokine activated killer cells and recombinant interleukin-2 to patients with metastatic cancer. N Engl J Med 1985; 313:1485.

12. Beckner SK, Longo DL. Media for LAK cell generation. Cancer Treat Rep 1986; 71:104.
13. Parkinson DR, Snydman DR, Weisfuse I, Werner B, Graham D, Will M, Marcus S, Boldt D, Doroshow JHM, Rayner A, Glover A, Fisher R. Hepatitis A (HAV) infection occurring following IL2/LAK cell therapy. Proc Am Soc Clin Oncol 1987; 6:234.
14. Yannelli JR, Mrowca-Bastin A, Jadus M. To the Editor. Cytotechnology 1988; 1:183.
15. Jadus MR, Thurman GB, Mrowca-Bastin A, Yannelli JR. The generation of human lymphokines activated killer cells in various serum free media. J Immunol Methods 1988; 109: 169.
16. Stevenson HC, Fauci AS. Counter current centrifugation elutriation. In Hewscowitz HB, Holden HT, Bellanti JA, Jhaffar A (eds): Manual of Macrophage Methodology. New York, Marcel Dekker, 1981:64–71.
17. Stevenson HC, Schlick E, Griffith R, Chirigos MA, Brown R, Conlon J, Kanapa DJ, Oldham RK, Miller PJ. Characterization of biological response modifier release by human monocytes cultured in suspension in serum-free medium. J Immunol Methods 1984; 70:245.
18. Shinomiya H, Nakano M, Shinomiya M, Miller PJ, Stevenson HC. Novel human monocyte-mediated tumor cell cytotoxicity assay under serum-free non-adherent conditions: Comparisons with conventional assay systems. (Submitted).

13

Clinical/Technical Challenges in Adoptive Cellular Immunotherapy: Role of Nurses, Technicians, and Other Support Personnel

MARGARET FARRELL, STEVEN RUDNICK, and PAUL J. MILLER
National Cancer Institute, Frederick, Maryland

In recent years there has been increasing interest focused on the "ex vivo" manipulation of leukocytes involved with immune system function that could offer major breakthroughs to the treatment of cancer. One new approach to treating metastatic cancer is adoptive cellular immunotherapy (ACI), a treatment in which activated antitumor leukocytes (that have been activated in the laboratory) are transferred to the cancer patient. The major obstacle to the use of ACI has been the inability to generate from cancer patients immune cells with antitumor reactivity in numbers sufficient for cancer treatment. In addition, the high doses of potentially toxic activating biological response modifiers (BRM), such as interleukin-2 (IL-2), that are required for lymphokine-activated killer (LAK) cell generation, pose very substantial patient management problems which affect every phase of current ACI trials (1).

With these ACI challenges in mind, the role of nurses, technicians, and other support personnel takes on added importance.

The theoretical concept of the term *role* may have several meanings; for the purposes of this chapter, role is defined as carrying out the rights, expectations, and obligations associated with a position. Nurses, laboratory technicians, and other personnel who may influence the well-being of the cancer patient are being challenged to identify their roles in the execution of ACI clinical research and treatment. When compared with standard expectations and behaviors for health care professionals, there are differences as well as similarities associated with the care of cancer patients receiving ACI that will be addressed in this chapter.

As a generalization, however, several introductory principles should be firmly kept in mind:

1. ACI represents the most complicated set of cancer therapies yet devised.
2. Certain ACI are as toxic as any cancer therapies yet devised.
3. In an attempt to raise efficacy rates and lower toxic side effects, certain ACI protocols undergo many week-to-week (sometimes day-to-day) revisions; knowledge updating is essential.
4. ACI are custom-made; that means that every patient is only reinfused with his own laboratory-activated immune system cells. Thus, each patient's therapy is truly a new experiment and, as such, is unique.
5. The uniqueness of each patient's ACI treatment (coupled with the seemingly endless series of steps required to complete each treatment) mandate that each member of the ACI team (physician/nurses/doctoral-level scientists/laboratory technicians/other support staff) be aware of his role in the overall therapy and be in open communication with the other members of the team.
6. The "front-line" professionals (nurses/technicians/other support staff) are just as likely to develop meaningful insights into the challenges and opportunities surrounding ACI as are the physicians and doctoral-level scientists. Be awake, be aware, be well-read, and (above all) *communicate* your observations and discoveries to others!

ROLE OF NURSES IN THE ACI TEAM

From the nursing perspective, six areas pertinent to caring for ACI patients will be discussed. These areas include (a) assessment; (b) emotional support; (c) cytapheresis, (d) administration of treatments, (e) management of adverse effects, and (f) documentation.

Assessment

A complete patient assessment (including an interview with the patient, family, other health care personnel; and a review of the medical record, laboratory studies, circulatory, respiratory, cardiac, and emotional status) are important to gather as baseline data. This information is essential in establishing nursing diagnoses and to plan, implement, and evaluate the patient's immediate care and long-term management.

A vascular access evaluation should be made to determine the possible need for placement of a venous access device. Other considerations at this time are the anticoagulant to be used during cytapheresis, the length of the cytapheresis procedure (2–5 hr), and the potential need for blood-warming equipment (or any other special equipment) to ensure the comfort and safety of the patients during the administration of ACI. Patients who are being considered for tumor-infiltrating lymphocyte (TIL) therapy, should be evaluated to determine if any contraindications or special considerations for surgery exist. Vital signs and nursing physical assessments should be performed to provide baseline data required to monitor the patient's clinical progress or to identify new problems (2).

Emotional Support

The psychological status of the prospective ACI patient merits special consideration, especially when formulating a patient management strategy; anxiety, stress, depression, fear, anger, and especially ignorance of the pertinent details of the ACI therapy can alter the effectiveness of the treatment. The patient's emotional status could greatly influence or impede the progress of care. Assessment of emotional and mental status includes the patient's level of consciousness, behavior and appearance, and intellectual function, including memory, knowledge, association, and judgment (3).

Patient education is important for patient compliance. The cancer patient receiving ACI needs a simple, concise explanation of the background, protocol design, and potential toxic effects of the treatment being offered. This process is important in investigational treatments because patients must decide if they want to participate in the protocols. Developing effective, consistent, and supportive patient education strategies is the joint responsibility of the physicians and the nurses involved in the study; such educational materials should include highly pictorial and written teaching aids, beyond the level provided in the informed consent.

A direct and honest approach is necessary to help the patient comprehend the management of his disease. Questions and verbalization of concerns should be encouraged. It is often beneficial if the nurse reviews the protocol design after the physician talks to the patient. Many times patients express fears and anxieties to the nurse that, for some reason, are not verbalized in the physician's presence. Consequently, the nurse assumes a pivotal part of the "information exchange" responsibilities in preparing the patient for the complex interventions yet to come. The ongoing teaching, information-sharing, and support provided by nurses help the patients deal with unrealistic expectations. Reassurance and practical advise is paramount to helping the patient during all phases of ACI therapy.

The nurse also needs to ensure that a signed informed consent has been obtained, and that the patient understands his decision. One of the nurse's important roles is to function as a patient advocate. The nursing role can thus extend into the role of instructor, counselor, or administrator; effective communication with the patient, family, and other health care professionals is paramount. Advocacy on the patient's behalf plus emotional support go hand-in-hand in the effective execution of ACI.

Gaining the patient's trust and developing a comfortable rapport also helps to reduce anxiety over a new and different treatment. This vital relationship can help support the patient who may be experiencing loss of control, not only because of the patient's entry into a new therapy environment, but also because of possible progression (or fears thereof) of the underlying malignancy.

Cytapheresis

Modern cytapheresis was originally designed as a therapeutic modality for certain diseases in which peripheral blood components play a role; this technique is now a critical first step in the performance of certain ACI protocols, such as LAK or AKM therapy. This procedure for harvesting large numbers of leukocytes by flow or filtration techniques is quite safe if strict aseptic methods are used to reduce the risk of infection. This section will address the degree of nursing involvement necessary to conduct safe ACI cytapheresis procedures (4).

The safe operation of a blood cell separator (cytapheresis device) requires a comprehensive understanding of basic cytapheresis principles, the details of operation of the specific brand of cytapheresis machine being employed by the institution, and extensive supervised "hands-on" experience. A comprehensive nursing/medical background is necessary to assess patient needs, monitor vital signs, and interpret this data in order to recognize the early stages of adverse reactions. Nurses are needed to safely administer blood and blood components, gain vascular access, administer drugs, and manage fluid volume during the cytaphersis procedure. The patient undergoing cytaphersis has many of the same needs as cancer patients receiving other complex investigational therapies.

Cytapheresis involves removing the patient's blood through a venous access catheter placed in the subclavian vein (or by using the patient's antecubital veins if accessible) and circulating the blood through a cell separator to remove the leukocytes. The remaining blood components (red blood cells and plasma) are then returned to the patient either by the venous access catheter or accessible peripheral veins through a large needle.

Safe, competent care calls for aseptic technique, careful attention to extracorporeal volume and replacement fluids, along with careful monitoring of blood counts during this procedure. Should any sign of distress become apparent, the procedure is stopped, an assessment made of the problem, and corrective action taken. Rapid access to a physician's bedside consultation and support is mandatory at all times.

The ACI cytapheresis procedure tends to average between 2 and 5 hr. Therefore, comfort of the patient during the procedure is important. Special attention to conditions such as warmth, chills, ventilation, position, hunger, thirst, and skin problems (itching) is helpful. Patients' comfort can be a critical determinant for how well they tolerate this phase of the ACI treatment overall. Moreover, the time spent in the cytapheresis unit is an excellent opportunity to develop the emotional and patient education rapport discussed in the preceding section.

Administration of Treatments

The nurse's role must include the safe administration of various therapeutic components pertinent to each ACI protocol. It is important that the nurse be clinically competent in administering the treatment, observing responses, and providing nursing interventions for the side effects and adverse reactions that may occur (5,6).

Once the leukocytes have been removed from the patient by cytapheresis, the pertinent leukocyte subset must be separated out and incubated with the appropriate biological response modifier (BRM) to generate activated cytotoxic leukocytes. After this laboratory incubation period, the activated leukocytes are then washed and resuspended in clinical-grade 0.9% saline and 5% serum albumin in preparation for reinfusion into the patient. The ACI can be administered through a venous access catheter for systemic therapy or (when indicated) by a more regional approach such as intra-arterial infusion for hepatic or renal cancer, intracerebral injections for brain tumors, or intraperitoneally for disease limited to the peritoneal cavity. The nurse must be familiar with the access devices for each of these ACI approaches, including long-term maintenance recommendations and procedures for sterile and safe entry into the catheter system employed.

Treatment administration for ACI patients must be standardized to ensure continuity and reproducibility of care. The cells are usually infused over 15 to 30 min. Gentle mixing of the cell suspensions is important to prevent clumping of the cells during the administration. The activated cell infusions must be given according to the schedule outlines in the treatment protocol. When follow-up BRM are administered (to maintain the cytotoxic activity of the infused cells), strict adherence to the schedule and administration technique outlined in the protocol is mandatory. Having ready access to a physician for bedside consultations, as required, is a prerequisite to the nursing administration of most current ACI treatments.

Management of Side Effects and Toxic Reactions

The goals of managing the side effects of ACI are to relieve the acute discomfort and prevent major side effects that can produce morbidity and mortality. Tables 1 through 3 outline the World Health Organization (WHO) classification and grading of toxicities observed in clinical trials.

General

Side effects commonly seen during ACI are fever, chills, and fatigue. The fatigue seems to worsen with long-term (> 6 months) treatments. Because patient comfort is a priority, these adverse reactions should be managed immediately during the course of treatment.

Management

The management of the fever, chills, and fatigue brought on by ACI include the following:

1. Fever should be monitored closely. Acetaminophen is given for temperatures above 101°F. If need be, cool sponge baths are given to control the fever. Cooling blankets may be necessary to control an excessively high temperature (> 103.5°F).
2. Chills can be managed by administering meperidine, 5 to 25 mg IV every 5 min until they have resolved. No more than 100 mg IV should be given within 60 min.
3. Fatigue can become a major side effect. Nursing care should be planned to conserve the patient's energy reserves. Assist the patient and suggest modification of life-style when indicated (e.g., frequent rest periods according to fatigue level).

Cardiovascular

The use of the BRM, interleukin-2 (IL-2), can cause cardiovascular side effects. An assessment of the heart and vascular system should be performed. This includes measuring the blood pressure and assessing the integrity of arteries and veins. Occasionally arrhythmias, angina, or ischemia are seen during this form of ACI. The arrhythmias are primarily atrial, although ventricular arrhythmias also have been seen. A frequent symptom observed is hypotension; this side effect with high-dose IL-2 is, at least in part, a result of the capillary leak syndrome (third-spacing) produced by this BRM. Third-spacing causes fluid retention, resulting in weight gain of up to 10 to 40 lb over the patient's baseline weight and a decrease in blood pressure and renal function. The nurse's observations are important to the subsequent design of the patient's diagnostic and therapeutic regimen.

Management

The management of cardiovascular symptoms that develop during ACI include the following:

1. The patient should be monitored for the development of arrhythmias. If noted, the physician should be contacted; continuation of therapy will depend on the severity and type of arrhythmias observed. If angina or myocardial in-

Table 1 World Health Organization Grading Classification of Cardiac, Neurological, and Pain Toxicities

	Grade 0	Grade 1	Grade 2	Grade 3	Grade 4
Cardiac					
rhythm	None	Sinus tachycardia >110 at rest	Unifocal PVC or atrial arryhthmia	Multifocal	Ventricular tachycardia or fibrillation
function	None	Asymptomatic but abnormal cardiac sign	Symptomatic dysfunction, no therapy required	Symptomatic dysfunction, responsive to therapy	Symptomatic dysfunction, unresponsive to therapy
Neurological					
level of consciousness	Alert	Transient lethargy	Somnolence <50% of waking hours	Somnolence >50% of waking hours	Coma
peripheral	None	Paresthesias	Severe paresthesias or mild weakness	Intolerable paresthesias or marked motor loss	Paralysis
constipation[a]	None	Mild	Moderate	Abdominal distension	Distension and vomiting
Pain[b]	None	Mild	Moderate	Severe	Intractable

[a]Constipation does not include constipation resulting from narcotics.
[b]Pain; only treatment-related pain is considered, not disease-related pain. The use of narcotics may be helpful in grading pain, depending upon the tolerance level of the patient.

Table 2 World Health Organization Grading Classification of Renal/Bladder, Pulmonary, Fever–Drug, Allergic, Cutaneous, and Infectious Toxicities

	Grade 0	Grade 1	Grade 2	Grade 3	Grade 4
Renal/bladder					
creatinine	$\leqslant$1.5	1.5–2.0	2.1–3.0	3.1–4.0	>4
proteinura	None	1^{+}, <0.1g/100 ml	$2\text{–}3^{+}$, 0.3–1.0g/100 ml	4^{+}, >1.0g/100 ml	Nephrotic syndrome
hematuria	None	Microscopic	Gross	Gross and clots	Obstructive uropathy
Pulmonary	None	Mild symptoms	Exertional dyspnea	Dyspnea at rest	Complete bedrest required
Fever–Drug	None	Fever <39°C	Fever 39–40°C	>40°C	Fever with hypotension
Allergic	None	Allergic edema, urticaria	Bronchospasm, no parenteral therapy needed	Bronchospasm, parenteral therapy required	Anaphylaxis
Cutaneous	None	Allergic edema, urticaria	Dry desquamation vesiculation	Moist desquamation, ulceration	Exfoliative dermatitis, necrosis requiring surgical intervention
Infections (specify site)	None	Infection requiring topical antibiotics	Infection requiring oral antibiotics	Infection requiring intravenous antibiotics	Infection with hypotension

Table 3 World Health Organization Grading Classification of Hematological and Gastrointestinal Toxicities

	Grade 0	Grade 1	Grade 2	Grade 3	Grade 4
Hematological					
hemoglobin (g/100 ml)	>11.0	9.5–10.9	8.0–9.4	6.5–7.9	<6.5
leukocytes ($1000/mm^3$)	>4.0	3.0–3.9	2.0–2.9	1.0–1.9	<1.0
granulocytes ($1000/mm^3$)	>2.0	1.5–1.9	1.0–1.4	0.5–0.9	<0.5
platelets ($1000/mm^3$)	>100	75–99	50–74	25–49	<25
hemorrhage	None	Petechiae	<50 ml blood loss	50–200 ml blood loss	>200 ml blood loss
Gastrointestinal					
bilirubin	⩽1.5	1.5–2.5	2.6–5.0	5.1–10	>10
SGOT/SGPT	<75	76–100	101–200	201–400	>400
Alkaline phosphatase	<150	150–200	201–400	401–800	>800
Oral	None	Soreness/erythema	Erythema, ulcers, can eat	Ulcers, requires liquid diet only	Cannot eat
Nausea/vomiting	None	Nausea	Vomiting requiring therapy	Intractable vomiting	Vomiting with dehydration
Diarrhea	None	<2 days	>2 days not requiring therapy	Intolerable requiring therapy	Hemorrhagic dehydration

farction is suspected, the infusion should be stopped and the side effects treated appropriately.
2. Patients who are hypotensive because of hypovolemia may require a fluid challenge; if no response to the fluid challenge is noted, vasopressors [such as phenylephrine (Neo-Synephrine) or dopamine] are usually given.
3. The third-space fluid retention usually resolves on its own when the IL-2 infusion is discontinued and the kidneys begin functioning properly. Diuretics may be necessary to relieve the patient's discomfort.

Respiratory

During LAK–IL-2 therapy the patient can develop respiratory distress (usually related to third-spacing from the capillary leak syndrome); the rate of breathing increases (tachypnea) up to 40 breaths/min. Each breath may become shallow and labored (dyspnea) and may be associated with grunting. Some patients develop rales and coughs. Patients may develop pulmonary edema.

Management

The principles of management of acute respiratory distress include the following:

1. Treat the cause if possible (such as tapping massive ascites and pleural effusions or reversing bronchospasm).
2. Maintain an open airway.
3. Administer oxygen as ordered and indicated by blood gas values (such as $pO_2 < 60$).
4. Monitor patient for signs of respiratory failure ($pCO_2 > 45$); intubation and mechanical respiratory support may be required.

Gastrointestinal

During LAK–IL-2 therapy the gastrointestinal side effects may be directly related to dose administration, but they do vary from patient to patient. Some patients are much more sensitive to specific biologicals, such as IL-2, than are others. Nausea, vomiting, diarrhea, and anorexia are common. With intraperitoneal (IP) therapy, the possibility of chemical peritonitis exists. This irritation of the peritoneal cavity may be due to IL-2 or lymphokine products induced by IL-2. Peritoneal fibrosis occurs in approximately one-third of the patients receiving IP LAK–IL-2, which makes continued treatment more difficult.

Management

The principles for gastrointestinal management of patients receiving IL-2 should include the following:

1. Assessment for evidence of generalized abdominal pain, rapid pulse, and evaluation of temperature.
2. Narcotics, if necessary, for the abdominal pain as ordered by a physician.
3. Intravenous fluid and electrolyte replacement therapy for the vomiting and diarrhea.
4. The nausea and vomiting can be minimized by using antiemetics such as prochlorperazine, as ordered.
5. For anorexia, utilize nutritional supplements if possible until the patient's appetite returns.
6. As yet, there is no effective prophylactic therapy against the emergence of peritoneal fibrosis, but a variety of approaches are being studied in the laboratory setting that will soon be attempted clinically.

It is important to note that these side effects usually resolve after IL-2 administration has been discontinued.

Genitourinary

The effect of IL-2 on renal function is one of the major toxic effects limiting a patient's ability to tolerate high doses of this BRM. Acute renal failure can develop during this therapy. Kidney function decreases in part because the vascular system is volume-depleted by the IL-2–induced capillary leak syndrome. Anuria is secondary to the fluid volume depletion and hypotension.

Management

To aid in the management of patients undergoing ACI the nurse must keep the following in mind.

1. Monitor the creatinine and BUN levels of the patient and relay any increase of these laboratory values to the physician.
2. Strict monitoring of intake and output is a necessity during LAK–IL-2 adoptive immunotherapy.
3. Low-dose dopamine to increase renal/arterial blood flow, and diuretics such as furosemide are often necessary to maintain urinary output.

It is important to reassure patients that their kidney function will return after the IL-2 administration is discontinued.

Neurological

Neurological assessments are imperative in patients receiving LAK–IL-2. Occasionally, patients experience headache, confusion, personality changes, and slowed neurological responses during the course of IL-2 administration. This may be the result of brain edema. Repeated dreams or nightmares have also been noted with this ACI.

Management

The objective of neurological toxicity management is to relieve the acute discomfort and avert danger by using the following methods:

1. Usually the headache attributed to IL-2 can be controlled with acetaminophen.
2. Watch for slowing of speech and a delay in response to verbal suggestions. These can be the early signs of IL-2–induced neurotoxicity.
3. Be alert for any sudden changes in the patient's condition, such as restlessness, disorientation, or increased drowsiness.
4. Reorient the patient frequently to assure the patient's recognition of his surroundings.

5. Ask family members to remain close by to offer the patient support and comfort.
6. Decrease external stimuli and noise for patient comfort.

All of these findings have neurological significance and must be documented and reported immediately to the physician.

A change in mental status can be very frightening for the patient and family. The nurse can help ease the family concern and the patient's difficulty with these side effects through optimism and reassurance that the symptoms will resolve after administration of IL-2 has been discontinued. The patient will require psychological assistance and encouragement while suffering these reactions.

Integumentary

Interleukin-2 causes erythema, pruritus, and even skin sloughing. These side effects again cause concern, aggravation, and decrease in body image. Chronic itching can decrease the patient's energy reserve and can even cause secondary infections (from open wounds).

Management

The major objectives of skin care therapy are to prevent damage to the healthy skin, prevent secondary infection, reverse the inflammatory process, and relieve the symptoms. Nursing interventions that help accomplish these objectives include the following:

1. Comfort measures such as cold, wet cloths to soothe the skin.
2. Oral or systemic medications, such as hydroxyzine, diphenhydramine, and the like, can be given to decrease the itching associated with the skin condition.
3. Many patients also find relief with lotions such as Lubriderm. Avoid preparations with alcohol or perfume content if possible.
4. Topical medications that suppress the inflammation, thus relieving the pain and itching, may be administered.

5. Bathing with Aveeno or even plain sitz baths have been helpful for some patients.

Musculoskeletal

Common musculoskeletal side effects seen with the administration of IL-2 are arthralgias. Occasionally, the arthralgias have caused extreme pain and disability.

Management

The musculoskeletal assessment consists of a general assessment of range and joint motion, muscle tone, and muscle strength. The management of the patient with musculoskeletal pain is important because the pain can be exhausting. The nursing interventions that may be helpful include the following:

1. Application of heat or cold may be beneficial in relieving muscle spasm and joint pain.
2. Analgesics and anti-inflammatory medications can be given per doctor's order for pain relief.
3. Sedatives before sleep have been helpful.

Hematological

Anemia and thrombocytopenia have been associated with LAK–IL-2; IL-2 directly inhibits hematopoiesis in these patients. The mechanism for these side effects are not yet completely understood. Patients with prior chemotherapy seem to experience thrombocytopenia to a greater degree. Eosinophilia also is a common side effect of IL-2 administration.

Management

It is important that the nurse monitor the patient's blood counts and alert the physician to any low counts.

1. Frequent monitoring of the blood counts
2. Replacement cell transfusions may be needed if:

   a. The hemoglobin falls below 8.0 g, or if symptoms related to low hemoglobin develop
   b. The platelet counts fall below 20,000, or if evidence o of bleeding problems related to low platelet levels develop
3. Patient education is required regarding:
   a. Easy bruising
   b. Bleeding gums
   c. Nose bleeds
   d. Petechiae
   e. Excessive fatigue

The eosinophilia usually is asymptomatic and resolves spontaneously when IL-2 is discontinued.

Documentation

Frequent clinical observation and physical assessment of the ACI patient can prevent serious reactions. Documentation should include treatment administration, treatment-related side effects, and pertinent diagnostic or laboratory tests. Treatment administration documentation should include the dose, route, site, time, and other medications given. The treatment-related side effects should be accurately described for type, quantity, and duration of the side effects noted. A flow sheet designed to highlight these and pertinent laboratory observations "at a glance" is quite helpful (8). The documentation should be objective, and the criteria measured by a grading system familiar to all those involved with the patient's care. By using a tool such as the World Health Organization Toxicity Grading Classification (see Tables 1–3) the nurse can easily assess for body system toxicities during the ACI treatment plus maintain standard evaluations useful in determining appropriate patient management strategies.

ROLE OF THE LABORATORY TECHNICIAN IN THE ACI TEAM

The laboratory technician who is in charge of purifying and activating the patient's leukocytes must not only have specialized expertise in this immunological technique, but must also be sensitive to a number of clinical concerns not usually faced by standard laboratory technicians. These include special attention to maintaining the cells sterile and free of toxins; expertise in sterile technique and the proper use of a laminar-flow hood is mandatory. Be certain that no nonclinical projects (like animal or radioactivity experiments) are performed in the designated ACI hood. The immunobiology technician must also be very sensitive to the handling and maintenance of any specialized apparatus that comes in contact with the patients blood: all equipment used in leukocyte isolation and culture must be sterilized or replaced with sterile components. When cells are washed and resuspended in preparation for patient-activated leukocyte administration, the technician must be mindful that these cells can only be resuspended in clinical-grade solutions; these solutions should be approved by the physician in charge of the ACI study, and fresh reagent stocks should be obtained on a regular basis from the hospital pharmacy. Take care not to use outdated solution stock. A log of the solution stock number and pertinent quality control data needs to be maintained on every patient sample. At a minimum, data for the microbiological cell culture sterility, viability, Gram stain results, cell counts, and in vitro cytotoxicity results should be maintained in log form.

Laboratory members of an ACI team should also bear in mind the unique needs of the clinical team and the patient. Open communication regarding the timing of each patient treatment is essential. The technician should be aware that punctuality is important; entire patient management plans revolve around the exact timing of each proposed activated leukocyte treatment. At the same time, the technician should realize that patient emergencies may come up at any moment, prompting the delay or cancellation of the proposed treatment. Again, excellent lines of communication are essential and technicians must be flexible in their ability to respond to the patient's needs and still maintain a high-quality activated leukocyte product. One mechanism to promote the sensitivity of laboratory members to patient clinical issues is to have laboratory representatives at pertinent patient rounds and discussions; this approach tends to enhance the awareness of laboratory members to the details of the status of patients that may affect future laboratory efforts. It is likely that national regulatory guidelines will soon be established to ensure the quality of the special breed of laboratory technicians who participate as members of the ACI team.

ROLE OF OTHER SUPPORT PROFESSIONALS IN THE ACI TEAM

Social workers, psychiatric liaison personnel, clinical coordinators, and dieticians all play important roles in the support of ACI patients. Each of these professionals should be aware of the general flow of the particular ACI treatment being administered, as well as the major adverse reactions expected. Support personnel should be encouraged to apply their specific skills to the details of each ACI patient; they should be provided with an open invitation to attend teaching and bedside rounds that pertain to their respective patients. Psychological and psychiatric staff should be aware, for example, of the potential neuropsychiatric complications of IL-2 or prolonged intensive care unit stay, and assist in dissecting ACI-related CNS impairments from reactive processes to the patient's malignancy or other factors. Social workers will be challenged when assisting families in securing financial compensation for these frequently experimental therapies. Dieticians will not only require skill in formulating diets for cancer patients but will also need to consider some of the unique effects of biological cancer therapies on appetite and protein metabolism. This information, coupled with an awareness of the stresses imposed on the patients and their families [by the underlying disease process (and possibly progression), the new research therapy enviroment, and the financial impact of treatment], will help to make all support personnel more knowledgeable, sympathetic, and effective.

CONCLUSIONS

As we have gained an increased understanding of the immune system's function in the emergence and therapy of the malignant process, we have witnessed a renewed enthusiasm for immunotherapeutic clinical trials in the 1980s. Recently, cancer biotherapy has been acknowledged as the fourth cancer treatment modality (along with surgery, radiation therapy, and chemotherapy). Biotherapy approaches may eventually replace other conventional cancer treatments as first-line therapies for such serious illnesses as lymphomas and hairy cell leukemia. However, it is clear that there are many limitations to cancer biotherapy at present; formidable theoretical and practical concerns exist for the application of this form of immunotherapy to many types of malignancy. It is clear that ACI research will continue to be the subject of intense investigation with T lymphocytes, monocytes, and natural killer (NK) cells. Many years of additional research will be required to devise mechanisms for increasing the clinical efficacy (as well as decreasing the toxicity) of this form of immunotherapy. The purpose of this chapter has been to detail the practical "front-line" approach to current ACI research and to outline the clinical challenges of this research arena from the nursing and technical perspectives. There is every reason to believe that new-generation ACI protocols will continue to grow out of the valuable observations and contributions of each ACI team member, including nursing, laboratory technician, and support personnel.

REFERENCES

1. Stevenson H C, Stevenson G W. Adoptive cellular immuno-

therapy. In Oldham R K (ed): Cancer Biotherapy: Principles and Practice. New York, Raven Press, 1987:385–397.

2. Mayer D K, Strohl R. Investigational cancer treatment modalities. Core Curriculum for Oncology Nursing. 1987: 237–244.
3. Hubbard S M, Seipp C A. Administering cancer treatment: The role of the oncology nurse. Hosp Pract 1985; 20: 167–174.
4. Passow J. Responsibilities of the registered nurse in the apheresis laboratory. Published by Mayo Clinic Blood Bank (Rochester, Minn), 1985.
5. Dewey D. Role of the nurse in the use of biological response modifiers. AAOHN J 1987; 35:163–167.
6. Sticklin L A. Interleukin-2 and killer T cells. Am J Nurs 1987; 4:468–469.
7. Brunner L S, Suddarth D S. Assessment of respiratory function. In Textbook of Medical-Surgical Nursing. 1980; 4:399–421.
8. Johnson B L, Gross J. Principles of clinical research. In Handbook of Oncology Nursing, 1985:67–92.
9. Morita T, Yonese Y, Minata N. In vivo distribution of recombinant interleukin-2–activated autologous lymphocytes administered by intraarterial infusion in patients with renal cell carcinoma. J Natl Cancer Inst 1987; 78:441–447.
10. Jacobs S, Wilson D J, Kornblith P L, Grimm E A. Interleukin-2 or autologous lymphokine-activated killer cell treatment of malignant glioma. Phase I trial. Cancer Res 1986; 46:2101–2104.

14

Adoptive Cellular Immunotherapy of Cancer: Future Perspectives

HENRY C. STEVENSON
National Cancer Institute, Bethesda, Maryland

Placed in perspective, adoptive cellular immunotherapy (ACI) has clearly been the one approach to cancer biotherapy that has generated the most interest and controversy in the past decade. The reasons for this are manifold. Certainly, all ACI protocols (as currently formulated) are time-consuming, labor-intensive, and costly; it has been estimated that each of these individual patient-customized therapies may cost as much as 50,000 dollars per patient, depending on the degree of support services required. The issue of ACI toxicity has also been of concern. Clearly, the lymphokine-activated killer lymphotyce (LAK) and tumor-infiltrating lymphocyte tumor-derived activated cell (TIL/DAC) protocols (as currently formulated) are as toxic as any cancer chemotherapies currently available; it appears that the biological response modifier (BRM), interleukin-2 (IL-2), is responsible for almost all of the side effects noted with these protocols. Perhaps identification of other BRM with less toxic side effects may make these two ACI therapies more tolerable. Also entering into the cost/benefit ratio controversy is the efficacy of ACI therapies. To date, only one ACI therapy, LAK–IL-2, has been subjected to formal multicenter Phase II (efficacy) trials; these studies indicate that between 15% and 25% is probably the average efficacy rate of LAK–IL-2 therapy as now formulated. Moreover, LAK–IL-2 does not appear to be active in all cancers; renal cell carcinomas, melanomas, and certain lymphomas appear to be most responsive. The LAK–IL-2 therapy does not now seem to substantially influence most malignancies (lung, breast, and gastrointestinal); moreover, a formal assessment of the affect of ACI therapy on cancer patient survival has not yet been performed.

Amid this controversy, one might wonder why the interest in ACI is so substantial. The reason is simple: if current results are confirmed (and improved) over time, they will form the first substantive evidence that the failing immune system of the cancer patient can indeed be "rewired" and "upregulated" in a fashion that allows the elimination of cancer cells from the body. This basic tenet of cancer biotherapy has heretofore been unproved. The early results of monoclonal antibodies in B-cell–derived malignancies, for example, have not been confirmed over time; nor have monoclonal antibodies been shown to have reproducible therapeutic effects on other malignancies. Certain BRMs (most notably the interferons) have been shown to induce substantial remissions in a variety of malignancies; most of these malignancies are leukocyte-derived and the mechanism of action is felt to be the direct antiproliferative effects of these BRMs on tumor cells that bear receptors for the BRM. It has not yet been shown, for example, that the high remission rate observed in hairy cell leukemia patients treated with interferon-α is in any way related to upregulation of any immune system component that subsequently eliminates the cancer cells. Instead, it appears that hairy cell leukemia cells are in some fashion directly downregulated by interferon-α itself by one of the many known antiproliferative effects of this BRM. Before ACI, no cancer biotherapy manipulation was able to reproducibly direct the immune system to eliminate cancer cells from the body. This ACI achievement (toxicity and low efficacy rates notwithstanding), represents a breakthrough discovery anticipated by immunology researchers for the past 100 years. A potential blemish on this achievement is the recent report (see Chap. 5) of response rates of IL2 alone that appear comparable to those observed with LAK + IL2. If substantiated, this new evidence will create new impetus for a precise determination of the mechanisms of action of LAK cells.

How will ACI develop in the years to come? The "big picture" answer to this question indicates that ACI (as currently performed) is simply an interim research biotherapy that will open up new horizons for less-expensive, less-toxic, less-time-consuming and, ultimately, more-efficacious cancer biotherapies. As shown in Figure 1, current Phase I and Phase II ACI trials are based on many years of preclinical observations in the ACI arena. As the limitations and toxic effects of ACI therapy are defined clinically,

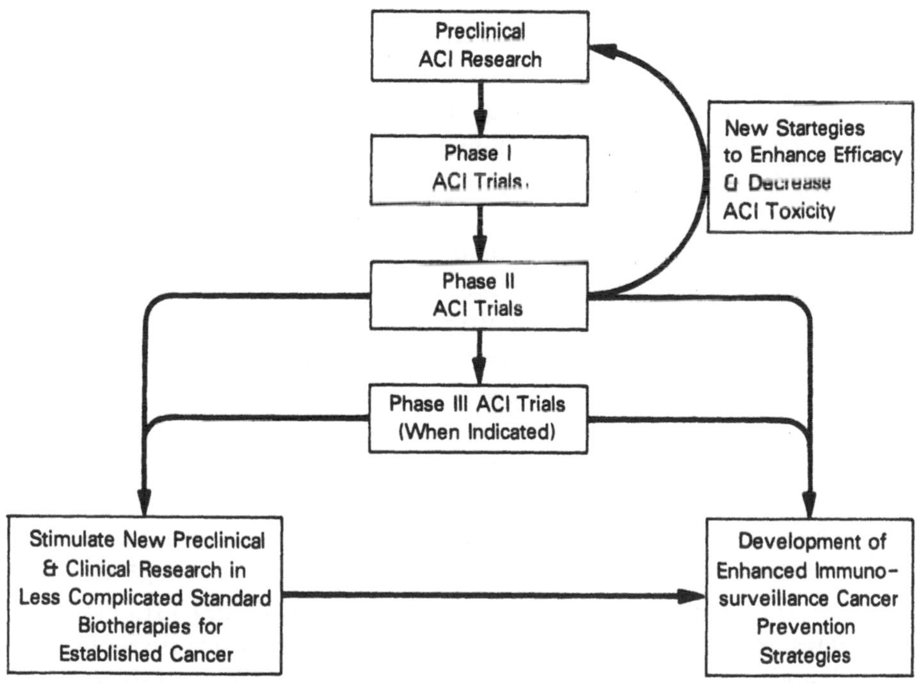

Figure 1 Diagrammatic representation of the likely evolution of ACI and its stimulatory influence on the development of newer standard biotherapies and ultimately enhanced cancer immunosurveillance initiatives.

this should promote feedback to the level of preclinical ACI research teams to help define newer approaches to the problem. At the same time, as ACI is shown to have some efficacy in the setting of refractory malignancies, an impetus is provided to develop less complicated standard biotherapies that will accomplish the same end-result as ACI. For example, if the mechanism of action of LAK–IL-2 therapy is found to be by production of high concentrations of IL-2 in the local vicinity of the tumor, perhaps other biotherapy strategies will be developed whereby high concentrations of IL-2 could be targeted to the site of tumor without the need for the cumbersome ACI approach. Finally, it should be kept in mind that just as ACI serves as an important probe for understanding the anticancer capabilities of the human immune system (that will allow us to develop more effective standard cancer biotherapies), both ACI and standard cancer biotherapy research will, in turn, point to new strategies whereby we can harness the immunosurveillance activity of the human immune system to prevent the emergence of clinical malignancy. We introduced this volume with a review of the current postulated mechanisms of action whereby the human immune system eliminates nonself invaders, including bacteria, viruses, and fungi (Chap. 1); the potential for this system to eliminate neoplastically transformed cells was also reviewed. It is entirely possible that the cumbersome, expensive, and frequently toxic biotherapies that ACI currently represents may ultimately be reformulated into the simple kinds of immunization/immunopotentiation strategies against cancer cells that we have routinely come to accept as effective defense mechanisms against other forms of nonself invaders (including the eradication of smallpox and, we hope, the eradication of AIDS).

The time course of the aforesaid evolution of ACI research is not entirely predictable. For the present, it is clear that ACI will undergo multiple permutations of the LAK, AKM, and TIL/TDAC themes utilizing many distinct BRM (and combinations thereof), with and without the addition of chemotherapy. We must exercise caution in the exploration of these avenues of inquiry so that we do not fall prey to the many of the imposing nonscientific pitfalls of ACI as currently performed. Substantial lessons can be learned from the history of chemotherapy developmental research and our recent experience in the development of effective standard cancer biotherapies. These two distinct modalities of cancer therapy have clearly demonstrated three basic clinical research principles that we must adhere to in ACI research. First, we must work with purified effector substances whenever possible. When impure chemotherapy agents were employed initially, it was difficult to determine which elements of the chemotherapy mixtures were producing beneficial effects and which ones were producing toxicity. This serious impurity problem posed ponderous impediments to the development of effective chemotherapy regimens. Once solved, the formal era of chemotherapeutic pharmaceutical agents was ushered in. The use of "natural" biological response modifiers (BRM) posed similar problems, in the early 1980s, to the development of new cancer biotherapies. Many of the natural BRM preparations studied at the time had less than 1% of the desired BRM protein: What was the other 99% and was it contributing to efficacy, toxicity, or both? The development of genetically engineered purified BRM has substantially improved our understanding of the physiology/kinetics of these agents and left us with a clear understanding of their toxicities and rates of efficacy. This breakthrough allowed cancer biotherapy to be declared the "fourth modality of cancer treatment" by 1985. No one performing cancer chemotherapy or cancer biotherapy research wants to return to an age of uncontrollable mixtures of anticancer agents.

With this principle in mind, every effort should be made to devise strategies whereby the effector cells in ACI are purified as much as possible and tested in formal Phase I/Phase II trial designs to understand the toxicities, efficacy, and physiology of each individually purified cytotoxic leukocyte subset. Secondly, combinations of anticancer agents should await the completion of Phase I/ Phase II testing of individual agents. This developmental model for combination cancer therapies has proved successful in the development of combination chemotherapies and is being closely adhered to in the development of combination biotherapies (combination of BRM). It is hoped that the development of "combination ACI" will await the development of strategies whereby these cytotoxic leukocyte combinations can be formulated in a rational and controllable fashion. As shown in Table 1, a final recommendation is that apheresis-related information be gleaned from Phase I and II trials of single purified BRM that are likely to be employed in future ACI research. The effects of these agents on the trafficking and sedimentation/separation properties of leukocyte subsets should be closely monitored. This information will be of immense benefit to the apheresis team in the design phases of future ACI that employ the previously characterized BRM. Figure 2 illustrates our current level of technology relative to ACI and the proposed future directions for developing this technology. The most solid scientific initial approach is to test the activity of purified subsets of cytotoxic leukocytes activated with a single BRM. Progression to use of controlled mixtures of subsets with single BRM and utilization of multiple BRM (plus/or minus chemotherapy) with purified subsets will certainly follow. In its most convoluted forms, ACI is likely to develop into formulations of controlled mixtures

Table 1 Guidelines for Researching ACI in a Scientific Fashion

1. Begin by administering purified activated leukocyte subtypes in order to characterize toxicity/efficacy/physiology of each.
2. Develop "combination ACI" only after each individual activated leukocyte subtype has been studied in Phase I/II totals. Create these combination ACI in a rational and precisely controllable fashion.
3. Monitor Phase I/II trials of purified individual BRM for changes in the trafficking pattern and sedimentation/separation properties of leukocyte subsets which may impact on the apheresis phase of future ACI.

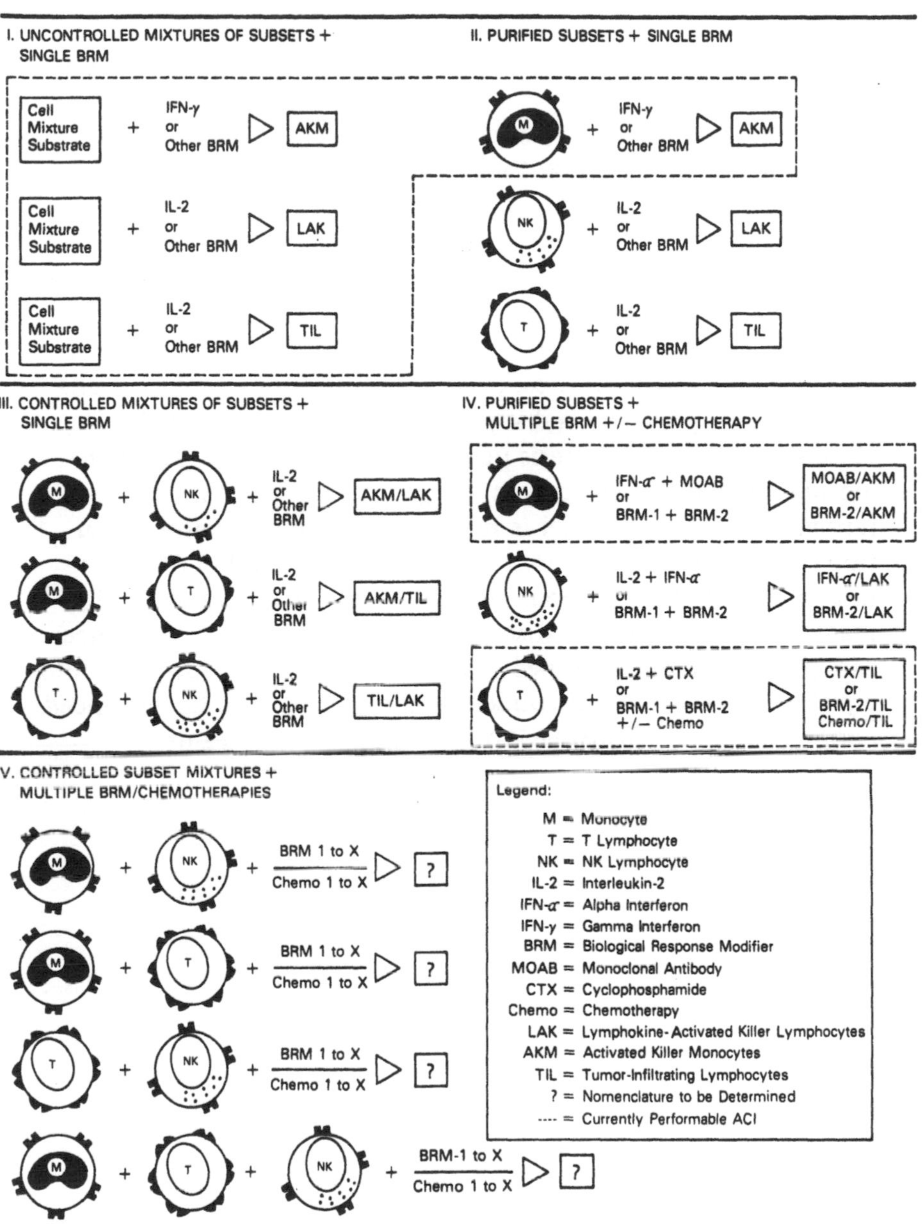

Figure 2 Current level of ACI technology and proposed future directions for its development.

of AKM, LAK, and TIL/TDAC, with multiple BRM and chemotherapies added at strategic points to maximize efficacy and minimize toxicity.

We may be able to look back from the heights of the 21st century and appreciate that the ACI trials being performed today were an integral part of a firm foundation for the development of simple and effective cancer biotherapies and cancer immunoprevention strategies. However, the excitement surrounding this historic effort should not obscure one other very important opportunity. This volume has stressed the multidisciplinary nature of ACI, because it requires the extensive cooperation of many teams of health care providers and support personnel professionals. The close association of apheresis personnel, biotherapy nurses, technicians, preclinical researchers, surgeons, oncologists, and intensive care unit personnel is required for the successful execution of a single ACI treatment. As we strive to develop ACI into new, more effective weapons against cancer, let us not waste this marvelous opportunity to discover more about ourselves as cancer care professionals. It is hoped that this volume has helped promote equally both the scientific understanding of ACI and the enhanced cooperation and effectiveness of participating ACI teams.

Index